Editor
Sara Connolly

Editorial Project Manager
Elizabeth Morris, Ph.D.

Editor-in-Chief
Sharon Coan, M.S. Ed.

Illustrator
Sue Fullam

Cover Artist
Brenda Di Antonis

Art Coordinator
Kevin Barnes

Art Director
CJae Froshay

Imaging
Craig Gunnell

Product Manager
Phil Garcia

Publishers
Rachelle Cracchiolo, M.S. Ed.
Mary Dupuy Smith, M.S. Ed.

Weekly Science Activities

Author

Amy Gammill, M.Ed.

Teacher Created Materials, Inc.
6421 Industry Way
Westminster, CA 92683
www.teachercreated.com
ISBN-0-7439-3842-9

©2004 Teacher Created Materials, Inc.
Made in U.S.A.

The classroom teacher may reproduce copies of materials in this book for classroom use only. The reproduction of any part for an entire school or school system is strictly prohibited. No part of this publication may be transmitted, stored, or recorded in any form without written permission from the publisher.

Table of Contents

Introduction
About the Lessons. 4
Finding Web Sites: Two Basic Strategies for Your Students 4
Lesson: Effective Internet Searches (Using a Search Engine) – Flower Parts*. 7
Lesson: Using a Subject Directory – Amazing Minerals*. 12

Earth Science
What Makes Up the Earth?*. 15
Changes in the Earth*. 17
Active Earth**. 19
Earthquake Patterns***. 22
Saving the Environment***. 26
Measuring Weather Conditions**. 29

Life Science
What's in a Name?**. 32
Life on Earth**. 35
Trees are More than Wood**. 37
Your Local Endangered Species**. 40
Classifying Living Things**. 43
Wildflowers in Your State*. 46
Ecosystem Research**. 49
Myrmecology: The Study of Ants**. 53
Insects*. 56
Vertebrates & Invertebrates*. 59
Animal Cam***. 63

Human Body
Healthy Food Habits*. 66
Cells: Body Building Blocks*. 69
Body Systems*. 71
Exercise Choices**. 73
Heredity: All About You***. 76

Table of Contents (cont.)

Science in History and Society

Science News** 80
Fields of Science** 83
Famous Aviators & Technology*** 85
Preventable Natural Hazards?*** 88
Changes in Communication*** 92

Space Science

Packing for Space* 96
International Space Station** 98
Phases of the Moon*** 100
Earth is Unique** 103
Our Solar System* 106
Constellations & Myths* 109
Lights in the Northern Sky*** 112

Science and Technology

Inventing the Flying Machine*** 114
Our World from Space*** 118
20th Century Inventions* 121
Science vs. Technology* 124

Physical Science

Electricity and Magnetism** 127
Newton's First Law of Motion** 129
Elements** 131
Chemistry in Action** 134
Forms of Energy*** 137
A Sun's Energy* 140

Difficulty key:
* * = elementary
* ** = intermediate
* *** = challenging

Introduction

About the Lessons

Most of the lessons in this book were written to be non-Web site specific. Students are instructed to use a search engine or subject directory to find a Web site that pertains to the topic of the lesson. To ensure student success, please read the section below and take the time to teach the first two lessons in this book, *Effective Internet Searches* and *Using a Subject Directory*. Please note that the author was careful to choose topics for which there is plenty of information on the Internet.

Suggested Web Sites

Note that all of the lessons use the Teacher Created Materials (TCM) Web site as a starting point for students. At the top of each student activity page, students are instructed to go to the TCM Web site and click a certain hyperlink (e.g., "Go to **http://www.teachercreated.com/books/3842** and click on page 96, site 1"). On the TCM Web site, students should click the link that corresponds to the lesson. Clicking the link will take them to the Web site they should use for the lesson. Most of the time, the link takes them to a search engine or subject directory. If you would prefer students go directly to the search engine or subject directory without first accessing the Teacher Created Materials Web site, by all means allow students to do so.

Alternate Web Sites

Just in case students are unable to find an appropriate Web site for a lesson, alternate sites are listed on the teacher page. Please note that at the time of this book's printing, these Web sites were working. It is possible that some Web sites will no longer be available when you try to access them. In this case, use a search engine or subject directory to find an alternate site for your students.

Difficulty Level

The lessons have been written at varying levels of difficulty. The following is a key:

* = elementary

** = intermediate

*** = challenging

Finding Web Sites: Two Basic Strategies for Your Students

It will be immensely helpful to your students to learn the following Internet search strategies – not only for the activities in this book, but also for any Internet research activity. Use the lessons *Effective Internet Searches* and *Using a Subject Directory* to teach these strategies.

Introduction

Strategy 1: Use a Search Engine

Search engines such as Google (**http://www.google.com**), Yahoo (**http://www.yahoo.com**), and Yahooligans (**http://www.yahooligans.com**) are tools that help users find information on the Internet. Type what you are looking for in the Search box on the site, press Enter, and a list of Web sites will appear. Click on a site name to go to a site and see if it is relevant to your topic. For more information on how to use a search engine effectively, see the lesson, *Effective Internet Searches*.

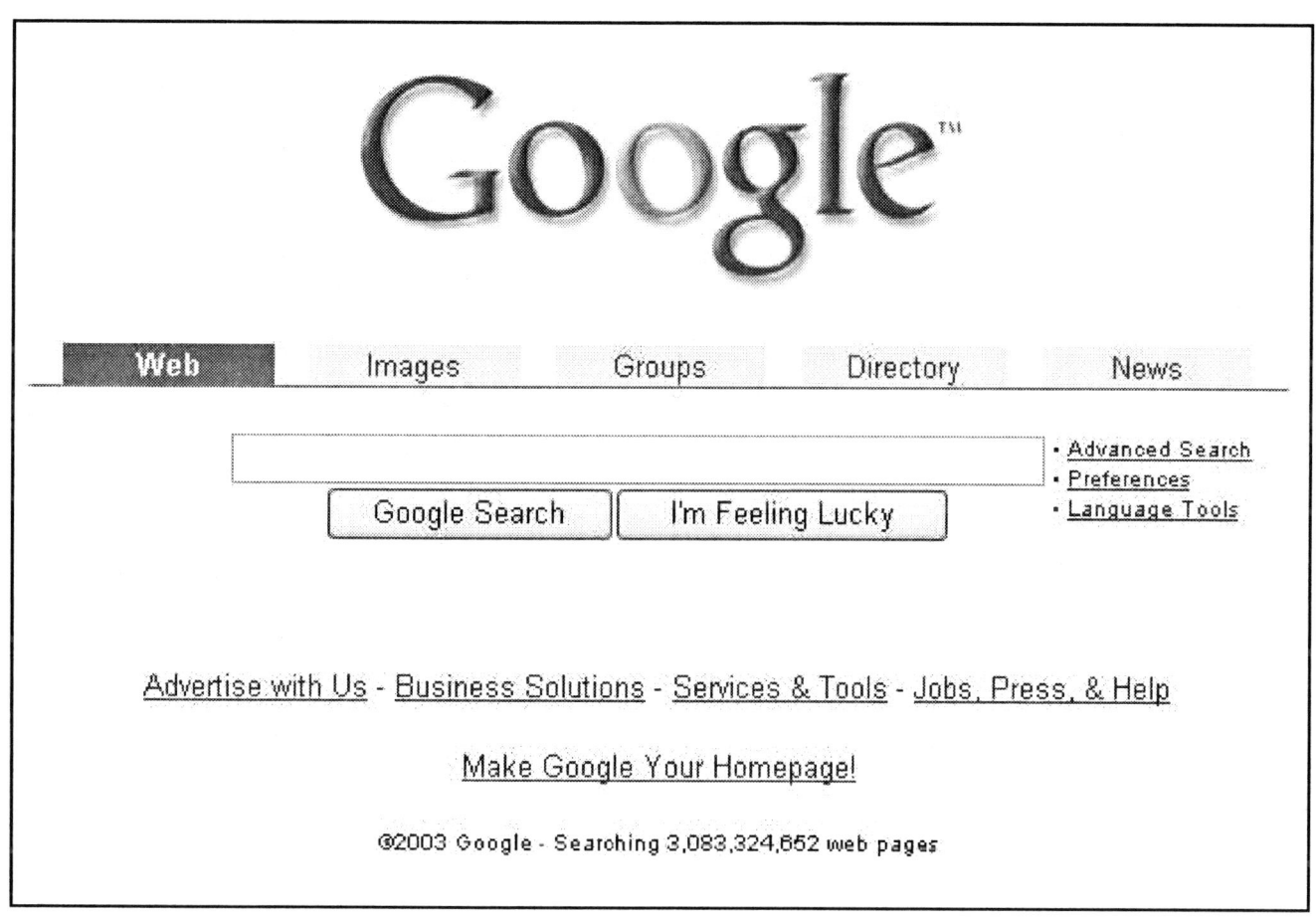

Google Home Page

When to Use a Search Engine

Search engines are best when you are looking for information on a specific person or term such as *Otto Lilienthal* (see the lesson, *Inventing the Flying Machine*), or when you have several criteria that can automatically narrow your search. For example, for the lesson, *Electricity and Magnetism*, students are instructed to type "scientists electricity magnetism" in the Search box to find information on a scientist who worked in the field of electricity and magnetism.

It takes some practice and a bit of know-how to make search engines most effective for you and your students' needs. Please use the first lesson in this book, *Effective Internet Searches*, to teach students to use search engines effectively. In addition, to ensure the best results in searching for information during each lesson, instruct students to use an *Effective Internet Searches Checklist* (see page 10) every time they perform a search using a search engine.

Introduction

Strategy 2: Use a Subject Directory

Web sites such as Yahooligans or KidsClick! contain subject directories that are more useful for finding information on a general topic. To use a subject directory, go to a site such as Yahooligans (**http://www.yahooligans.com**), click a subject area (such as **Science and Nature**), then a subtopic (such as **Astronomy and Space**), then the next subtopic (such as **Solar System**), and then your specific topic (such as **Sun**). (Note: Yahooligans and KidsClick! were created specifically for kids. For the most part, both sites will direct students to Web sites that were also created specifically for kids.) For more information on how to use a subject directory, see the lesson *Using a Subject Directory*.

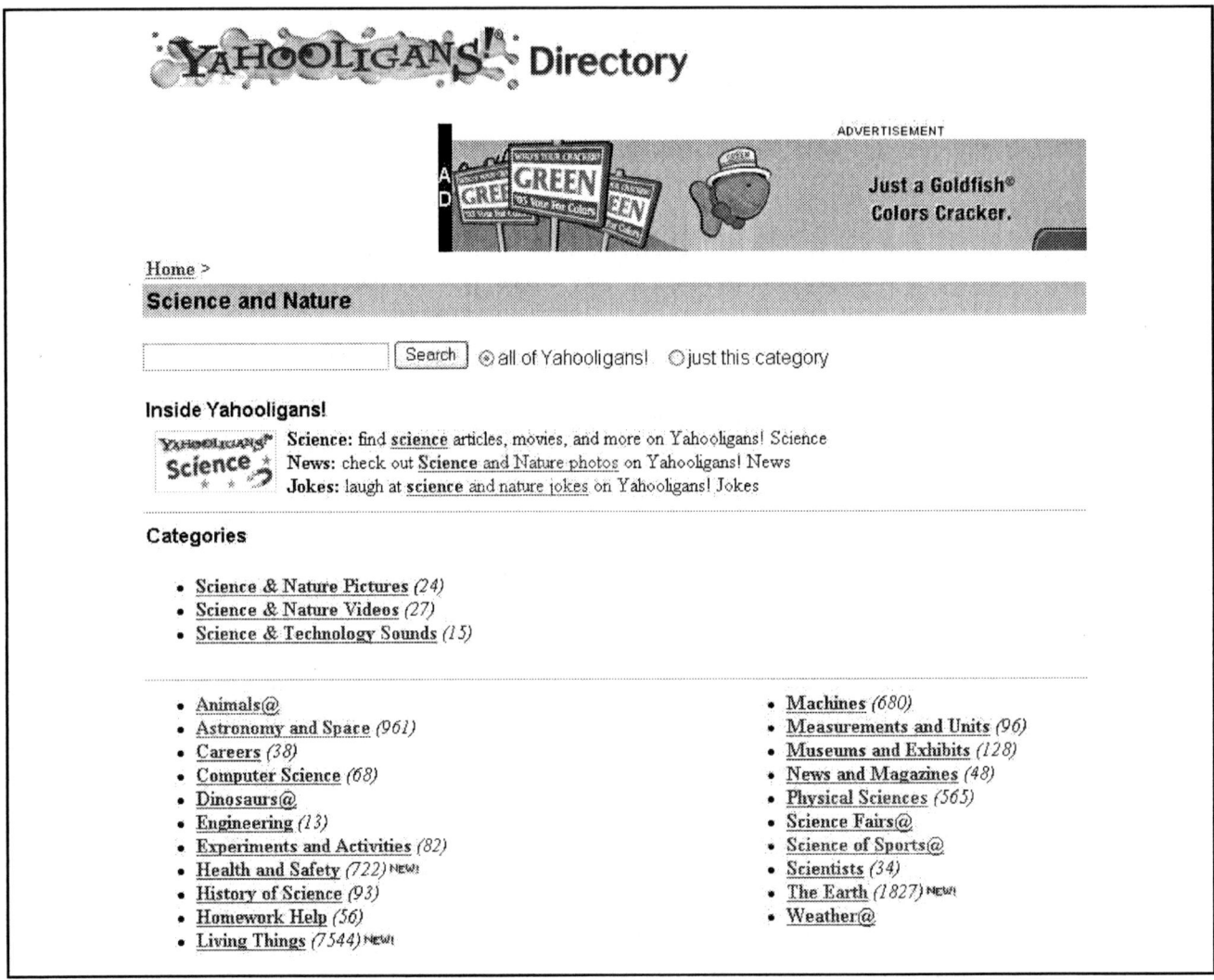

Yahooligans Subject Directory

When to Use a Subject Directory

Subject directories are best for finding information on a general topic such as the Sun (see the lesson, *A Sun's Energy*), frogs (see the lesson, *Classifying Living Things*), or human body systems (see the lesson, *Body Systems*). Please use the lesson titled, *Using a Subject Directory* to teach your students how to use a subject directory.

Introduction: Teacher Notes

Effective Internet Searches*
(Using a Search Engine)

NSTA Standards (5–8):
A, C

Objectives:
Students will:
- use a high-quality search engine
- generate specific and precise keywords
- conduct an Internet search using a search engine
- evaluate several Web sites for content-appropriateness, age-appropriateness, and organization/clarity
- open a new browser window
- choose the best Web sites for finding the information they need

Suggested Web Site:
http://www.teachercreated.com/books/3842

Click page 7, site 1

Alternate Web Site(s):
Yahooligans

http://www.yahooligans.com

Ask Jeeves for Kids

http://www.ajkids.com

Pre-Internet Activities:
- Ask students how they find information on the Internet. Allow students to share their ideas, including using search engines. Ask students to share what they know about search engines.
- Gather students around a computer or computer projection screen and show them a search engine (see suggested and alternate Web sites above). Tell students that today they will learn about the parts of a flower. Look at the Flower Parts student activity sheet together. Tell students that it is important to know what you are looking for before you start a search. Ask students to describe what they will need to find to complete the student activity sheet.
- Now ask for suggestions on what to type in the Search box to find information on the Internet about flower parts. Tell students they could type, "parts of a flower," "parts flower," or "flower parts." Ask which one is best. Point out that the last two suggestions are good because it is not necessary to type words such as "of, a, the, is," etc. into the Search box. However, the last suggestion, "flower parts" is best because it is stating exactly what we want to find: flower parts. (Note: word order does make a difference. It is best to type the keywords in the order you hope they will appear on a Web site.)

Introduction: Teacher Notes

Effective Internet Searches *(cont.)*

- Also point out that by typing "flower parts" no extra words have been used. Tell students that search engines work by searching for Web sites that contain all of the words you type. You will get the best results if you type ONLY and ALL the words you want to appear on a Web page that might contain the information you need. If we had typed, "flower parts green," for example, the search engine will look for sites that contain all of those words. Since the word green is not important for what we are trying to find, it is not useful to tell the search engine to look for it.

- Show students how to type the keywords "flower parts" into the Search box and press Enter on the keyboard. A list of Web sites will appear. Read the Web site titles and excerpts of the first several Web sites on the list.

Introduction: Teacher Notes

Effective Internet Searches *(cont.)*

Ask students how they might choose a particular Web site to view. First, looking at each Web site title and description in turn, ask the question, does the Web site appear to contain the information I need? Second, ask the question, does the Web site appear to be written for kids? Using these criteria, decide as a group which Web site is best for finding the names of flower parts. Then show students how to open a new browser window by clicking **File**, then **New**, and then **Window**. (Opening a new window will ensure the search results will be readily accessible to students when they want to go back to them.) Then click the title of the Web site you chose to view it.

Take a brief look at the Web site and instruct the students to ask themselves the following questions:

- Does this Web site appear to have the information I need?
- Is this Web site written for kids?
- Does this Web site appear to be well-organized and easy to read?
- Is it easy to move around this Web site (important concepts or subtopics are hyperlinked)?

   After briefly evaluating this Web site go back to the search results by clicking the tab on the task bar at the bottom of the screen. (The tab will say something like, **Google Search: flower parts**). Open a new window, choose a different Web site, and ask the same questions to evaluate it. Do this again with one more Web site and ask students which one they should use to complete the student activity sheet.

   Now allow students to go to their computers, perform a search using a search engine, and complete the Flower Parts student activity sheet. Students should also complete an *Effective Internet Searches Checklist* (found on the next page). Note: for best results with all the lessons, direct students to use an *Effective Internet Searches Checklist* every time they use a search engine to find information.

Extensions:

- Introduce students to the alternate search engines listed above. Sometimes these search engines will be more appropriate for your students because they are designed specifically to find sites that are written for kids (however, the drawback is a more limited choice of Web sites). Note that Ask Jeeves for Kids works a little differently than other search engines. Students are instructed to type a question in the Search box such as, "What are the parts of a flower?" Ask Jeeves then shows students a short list of "answers." Students must read each answer to determine which one is likely to help them.
- Occasionally you or your students will want to find a specific term or phrase on a Web page. Instead of reading the entire page to find it, simply press **CTRL** and then the **F** key on your keyboard, type your word or phrase in the **Find what** box, and press **Enter**. The word or phrase will be highlighted on the Web page. Note that this feature searches only the current Web page you are viewing. It does not search the entire Web site.

Introduction: Student Activity Sheet

Effective Internet Searches CHECKLIST

Name: _____ Date: _____ Period: _____

STOP and think! This checklist will help you use a search engine effectively. However, if you are researching a general topic like the Sun, frogs, or human body systems, try using a subject directory instead.

Directions: Use the checklist below to guide you when using a search engine to find information. Or, use it after you have searched to evaluate how you did.

The Search Engine
- ❏ I used a top-notch search engine like Google, Yahoo, or Yahooligans.

Keywords
- ❏ I looked at the student activity sheet to know what I need to find.
- ❏ I thought about what I am looking for in terms of specific and precise words. I wrote down the keywords I should use here:

  _____

- ❏ In the Search box I typed ONLY and ALL the keywords that should appear on the Web page I am looking for.
- ❏ I DID NOT type extra words like "and," "the," "is," etc.
- ❏ I spelled the keywords correctly.

Choosing the Best Web Site
- ❏ After pressing Enter, I did not automatically choose the first Web site listed. I read the titles and excerpts of the first several Web sites on the list.
- ❏ I opened a new browser window before I clicked one of the Web sites listed.
- ❏ I viewed a few different Web sites before deciding which one(s) would give me the information I need.
- ❏ The Web site I chose is well-organized and easy to read.
- ❏ The Web site I chose is easy to move around.
- ❏ The Web site I chose was written for kids.

Introduction: Student Activity Sheet

Effective Internet Searches (cont.)

Name: _____ Date: _____ Period: _____

Directions: Go to **http://www.teachercreated.com/books/3842** and click page 7, site 1. Use the search engine to find a Web site that will help you label the flower parts shown below. To find a Web site that will help you, type the keywords, "flower parts" in the search engine Search box and press Enter on your keyboard. Read the titles and excerpts of the first few Web sites on the list, view a few of them, and then select the best one. After you are finished labeling the flower, use the Effective Internet Searches Checklist to evaluate how well you used the search engine.

Flower Parts

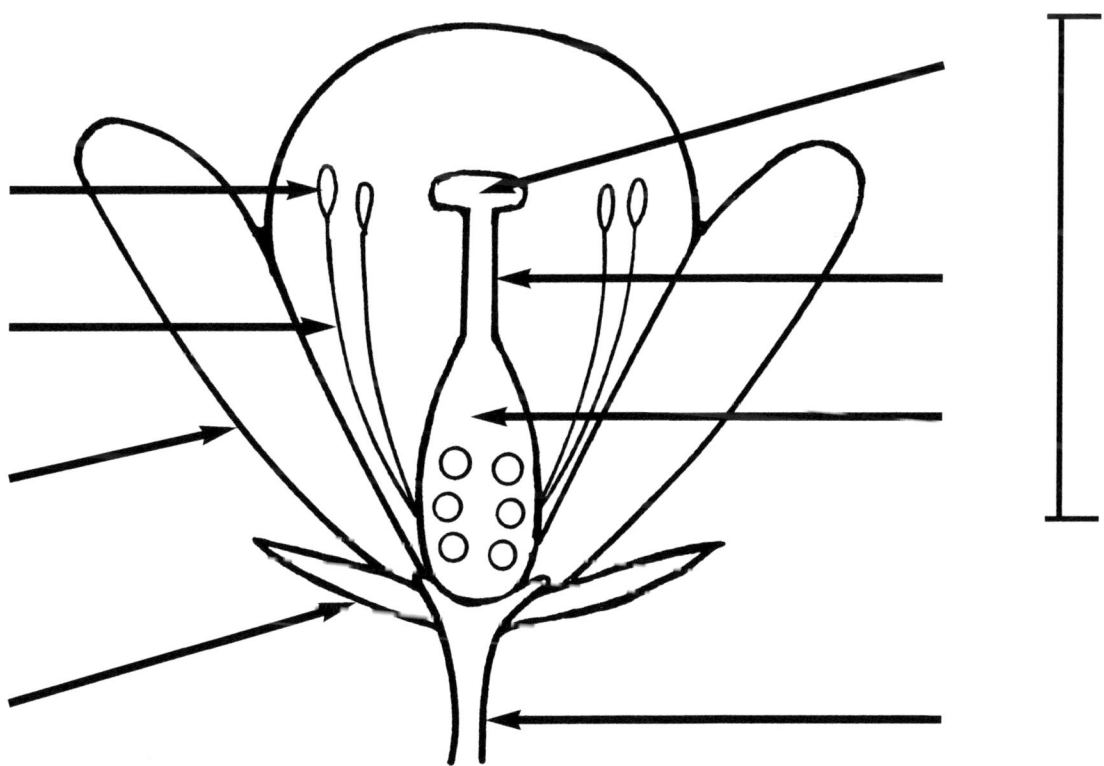

©Teacher Created Materials, Inc.   11   TCM# 3842 Web Resources for Science Activities

Introduction: Teacher Notes

Using a Subject Directory*

NSTA Standards (5–8):
A, D

Objectives:

Students will:

- visualize subject directories in terms of an inverted triangle
- discuss where to find subject directories on the Internet
- navigate a subject directory to find information
- use a subject directory to learn about minerals

Suggested Web Site:

http://www.teachercreated.com/books/3842

Click page 12, site 1

Alternate Web Site(s):

KidsClick! http://sunsite.berkeley.edu/KidsClick!/

Pre-Internet Activities:

- Brainstorm a sample subject directory as a class on a science topic of your choice. As students brainstorm, draw the directory path on the board in two ways: in a linear fashion with arrows, and as an inverted triangle. Here is a sample directory path drawn both ways:

    Science → Biology → Animals → Mammals → Lions

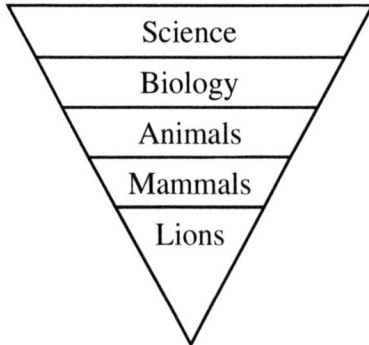

- Tell students that on certain Web sites there are directories just like the one they brainstormed. These directories help Internet users find Web sites on a general topic such as lions.
- With students gathered around the computer or projection screen, go to the suggested Web site and show them how to click through a directory to find Web sites on a general topic. For example, you might first click Science and Nature, then Physical Sciences, then Energy, then Fossil Fuels to see a list of Web sites on fossil fuels. Click through a few more directories to get a sense of the breadth of the offerings on the Web site.
- Allow students to complete the Rocks and Minerals student activity sheet using the suggested Web site's subject directory. Direct students to also complete the inverted triangle student activity sheet to document the subject directory path they took to find information on rocks and minerals.

Introduction: Student Activity Sheet

Using a Subject Directory *(cont.)*

Name: _____ Date: _____ Period: _____

Directions: On the inverted triangle below, record the subject directory path that led you to the information you sought.

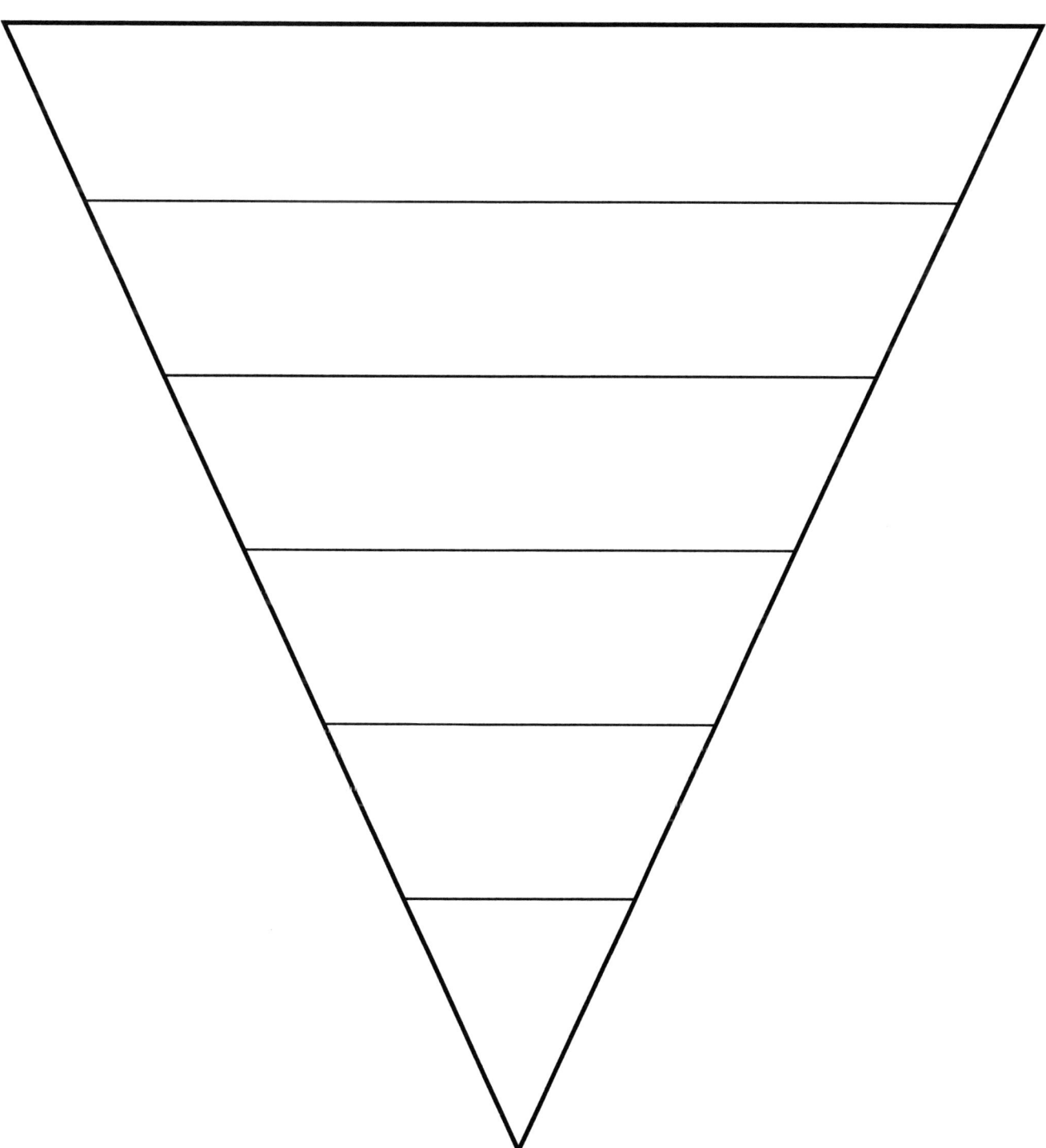

Introduction: Student Activity Sheet

Using a Subject Directory (cont.)

Name: _____ Date: _____ Period: _____

Directions: Go to **http://www.teachercreated.com/books/3842** and click page 12, site 1. Use the subject directory to complete this page. Hint: Rocks and Minerals are a subtopic of Geology, which is a subtopic of Earth Science.

Amazing Minerals

Minerals can be described and classified using several properties. Here are descriptions of four of these properties:

Color: The color of the mineral.
Luster: The way the mineral reflects light. Minerals can have metallic or non-metallic luster.
Streak: The color of the mineral when it is grounded into a powder.
Hardness: How hard a mineral is based on what it can scratch and what scratches it.

Now using the Internet, name five minerals and describe one property of each (such as color, luster, streak, or hardness). The first one has been done for you.

Mineral Name	Property	Description
Gold	Streak	Yellow to orange color

Bonus question: What is the difference between rocks and minerals?

Earth Science: Teacher Notes

What Makes Up the Earth?*

NSTA Standards (5–8):
A, D

Objectives:
Students will:
- identify and label the layers of the earth
- find and record one fact about each layer
- understand the general composition of the layers of the earth

Suggested Web Site:
http://www.teachercreated.com/books/3842

Click page 15, site 1

Alternate Web Site(s):
Layers of the Earth
http://www.svsd.sk.ca/grassroots/2001/project2/proj4/earth_gr3/layersoftheearth.html

Pre-Internet Activities:
- Ask students the following: if it were possible to slice the earth in half, what would they expect to find inside? Explain that scientists think the earth is made up of three major layers, each one being distinctly different than the others. We live on the outermost layer of the earth.

Post-Internet Activities:
- After students have found the names (crust, mantle, and core) and one fact about each of the main layers of the earth, briefly describe the composition of each layer. Explain that the crust is made up of solid moving plates that "float" on top of the molten mantle. The mantle is very hot and liquid. This is where lava comes from. The core of the earth is made up of dense heavy metals – nickel and iron. The outer core is liquid nickel and iron, and the inner core is solid nickel and iron.

Extensions:
- Direct students to brainstorm how scientists are able to determine what is inside the earth without actually going there. After brainstorming and sharing ideas, direct students to form a hypothesis. Remind students that scientists might use a variety of techniques or information to determine what is inside the earth. Then invite students to find the answer on the Internet. The suggested Web site is a good place to start.
- Using this lesson as a starting point, continue on to study the tectonic plates of the lithosphere. Discuss how the movement of the plates is the reason why there are earthquakes, volcanic eruptions, and mountains. Go even further and have students plot earthquake or volcano locations using data from the Internet (see the lesson titled, *Earthquake Patterns*), and then have students look for patterns. The patterns will correspond to the plate boundaries where plates collide.

Earth Science: Student Activity Sheet

What Makes Up the Earth? *(cont.)*

Name: _____ Date: _____ Period: _____

Directions: Go to **http://www.teachercreated.com/books/3842** and click page 15, site 1. Use the search engine to find the names of the three major layers of the earth, as well as one fact about each layer. Type the keywords "earth's layers" in the Search box and press Enter. Select the site with the answer that seems most appropriate.

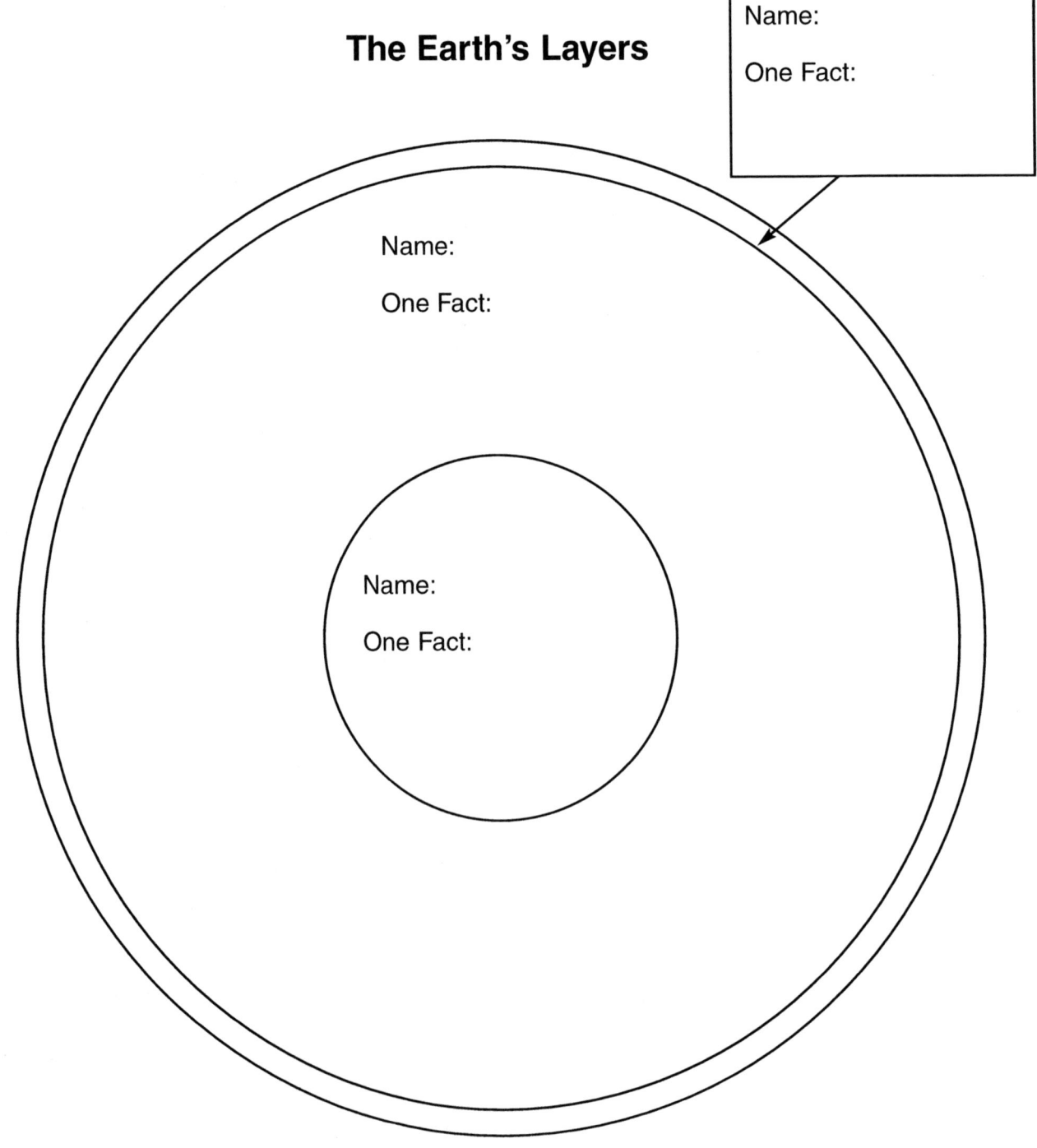

The Earth's Layers

TCM# 3842 Web Resources for Science Activities ©Teacher Created Materials, Inc.

Earth Science: Teacher Notes

Changes in the Earth*

NSTA Standards (5–8):
A, D

Objectives:
Students will:
- match geological vocabulary words to their correct definitions using information from the Internet
- create a word puzzle for several more geological terms

Suggested Web Sites:
http://www.teachercreated.com/books/3842
Click page 17, sites 1, 2

Alternate Web Site(s):
ZDNet Downloads
This is a shareware site, meaning you can download the software programs for free. On the homepage, type "word puzzles" in the Search box and press Enter. This will generate a list of downloadable word puzzle software programs that your students can use.
http://downloads-zdnet.com.com/?legacy=cnet
Make Your Own Word Search Puzzle
http://www.armoredpenguin.com/wordsearch/
Free Online Puzzle Maker
http://www.puzzle-maker.com/

Pre-Internet Activities:
- This activity is best done to reinforce terms to which students have already been introduced. Before doing this activity, be sure to discuss major geological concepts such as the layers of the earth, weathering and erosion, tectonic plates, earthquakes, and volcanoes.
- Show students how to create a word puzzle using the suggested Web site or an alternate Web site. Note: the Web site you choose to use may allow students to create a variety of different word puzzles such as crossword puzzles or word finds.

Extensions:
- After creating the word puzzles, allow students to print them and trade them with other students to solve each other's puzzles. If you do not wish to print the puzzles, allow students to trade computers to solve each other's puzzles.

List of Vocabulary Words for Creating Word Puzzles

geosphere (crust, mantle, core)	mantle	rock cycle
volcano	core	metamorphic rocks
lithosphere	sediment	igneous rocks
crust	weathering	sedimentary rocks

Earth Science: Student Activity Page

Changes in the Earth *(cont.)*

Directions: Go to **http://www.teachercreated.com/books/3842** and click page 17, site 1 to match the geological terms found below to their correct definitions. At the Web site, type each word in the Question box, press Enter, and choose the answer that is most useful for finding the definition. Once you have found the definition, write the number of the word's definition next to the word.

Changes in the Earth

Words   **Definitions**

_____ lava   1. to burst out

_____ magma   2. the process of wearing down or breaking apart rocks

_____ tectonic plates   3. hot, melted rock deep inside the earth

_____ erosion   4. a shaking or sliding of the earth's crust

_____ erupt   5. the rigid moving pieces of the earth's crust

_____ earthquake   6. hot, melted rock that flows from a volcano

---

Directions: Now go to **http://www.teachercreated.com/books/3842** and click page 17, site 2 to create a word puzzle. After creating your own word puzzle online, print it or write it in the space provided below.

Earth Science: Teacher Notes

Active Earth**

NSTA Standards (5–8):
A, D, F

Objectives:
Students will:
- understand that the Earth's surface is broken into several major plates which float on the mantle
- understand that at plate boundaries there are faults of three major types: divergent, convergent, and transform.
- determine the tectonic plate names, types of faults, major geological events caused by plate movement, and other interesting facts about plate tectonics using the Internet
- draw the three major types of faults and show plate direction with arrows

Suggested Web Site:
http://www.teachercreated.com/books/3842

Click page 19, site 1

Alternate Web Site(s):
Earth's Continental Plates
http://www.enchantedlearning.com/subjects/astronomy/planets/earth/Continents.shtml

Google
http://www.google.com (search for "plate tectonics")

Pre-Internet Activities:
- Explain to students the theory of plate tectonics. Describe how the Earth's surface, called the lithosphere or crust, is broken up into several major plates that slide on top of the hot, flowing mantle underneath. The plates move independently of each other, sometimes colliding, sometimes moving apart, and sometimes sliding against each other in opposite directions. The boundaries of the plates, where the plates meet, are called faults. There are three major types of faults: divergent faults, convergent faults, and transform faults. Plate movement causes major geologic events such as volcanic eruptions and earthquakes.

Extensions:
- Allow students to find actual examples of each of the three major types of faults. For example, the Mid-Atlantic ridge is an example of a divergent fault.
- To tie this lesson to the *Our World from Space* lesson, ask students to think about how satellites can help scientists confirm the theory of plate tectonics. Specifically, what type of satellite data would scientists find most useful for studying this phenomenon?
- Teach the *Earthquake Patterns* lesson after this one to help students understand why earthquake locations follow patterns.

Earth Science: Student Activity Page

Active Earth (cont.)

Name: _____ Date: _____ Period: _____

Directions: Go to **http://www.teachercreated.com/books/3842** and click page 19, site 1. Use the Web site to complete the concept map below. For example, in one of the spaces branching out from the category "Major Geological Events Caused by Moving Plates," you could write, "volcanoes."

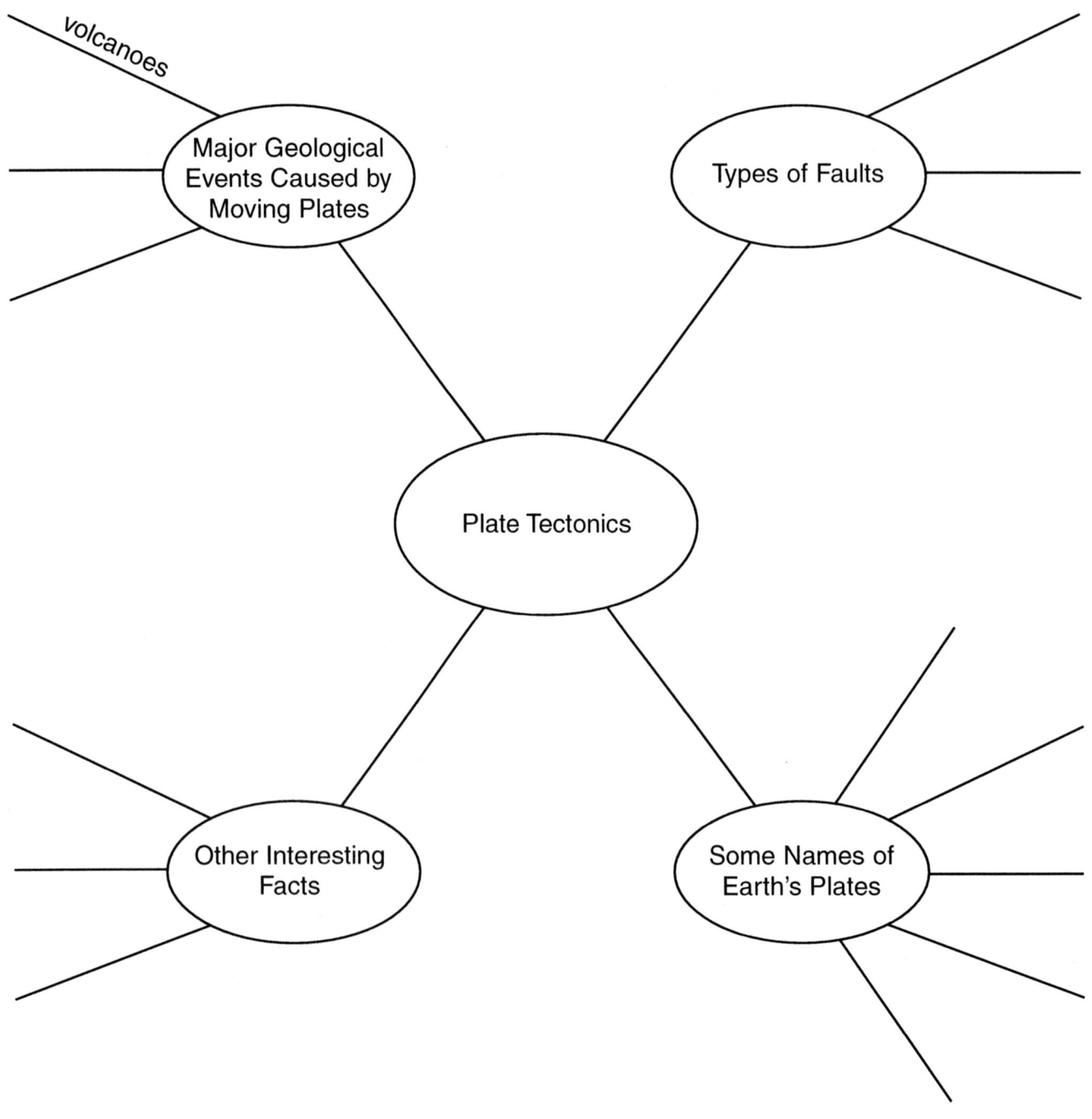

Earth Science: Student Activity Page

Active Earth (cont.)

Name: _____ Date: _____ Period: _____

Directions: Use the same Web site (go to **http://www.teachercreated.com/books/3842** and click page 19, site 1) to find and draw the three major types of faults in the boxes below. Be sure to show with arrows the directions in which the plates are moving.

Divergent Fault

Convergent Fault

Transform Fault

Earth Science: Teacher Notes

Earthquake Patterns***

NSTA Standards (5–8):
A, D, F

Objectives:
Students will:
- brainstorm what they know and want to learn about earthquakes
- plot the locations of earthquakes on a map using latitude and longitude data
- analyze the earthquake locations to find patterns
- describe the connection between earthquake locations and plate tectonics

Suggested Web Site:
http://www.teachercreated.com/books/3842

Click page 22, site 1

Alternate Web Site(s):
Google

http://www.google.com

(Use the keywords "earthquake locations United States")

USGS Earthquake Hazards Program – For Kids Only
http://earthquake.usgs.gov/4kids/

Pre-Internet Activities:
- To supply your students with some prior knowledge before this lesson, teach the lesson *Active Earth*. You may also wish to further explore the connection between plate movement and earthquakes before beginning the lesson.
- As a class or individually, direct students in completing the first two columns of the KWL chart. Note: save the third column titled, "What have we Learned about earthquakes?" for a post-lesson activity.
- Provide colored pencils to students to plot the earthquake locations. If you wish to add a level of complexity, have students use different colors for the varying magnitudes of the earthquakes. Ask them to create a key with magnitude ranges. For example, the color yellow might represent earthquakes with a magnitude of zero to 1, green might represent earthquakes with magnitudes ranging from 1 to 3, and so on.
- You may want students to print out the list of earthquake locations and circle the ones they plot. This way you can do a quick spot-check of their work.

Extensions:
- Direct students to explore how scientists pinpoint the epicenters of earthquakes and how they measure their magnitudes. Plenty of information is available on the Internet regarding these and other earthquake-related topics.

Earth Science: Student Activity Page

Earthquake Patterns *(cont.)*

Name: _____ Date: _____ Period: _____

Directions: Use the KWL chart below to brainstorm what you know and want to learn about earthquakes. Complete the last column titled, "What have we Learned about earthquakes?" after you have finished the earthquake location plotting activity.

What do we **Know** about earthquakes?	What do we **Want** to know about earthquakes?	What have we **Learned** about earthquakes?

Earth Science: Student Activity Page

Earthquake Patterns *(cont.)*

Name: _____ Date: _____ Period: _____

Directions: Go to **http://www.teachercreated.com/books/3842** and click page 22, site 1. Use the data found on this Web site to plot on the map below the locations of at least 20 recent earthquakes. Then answer the questions on the next page.

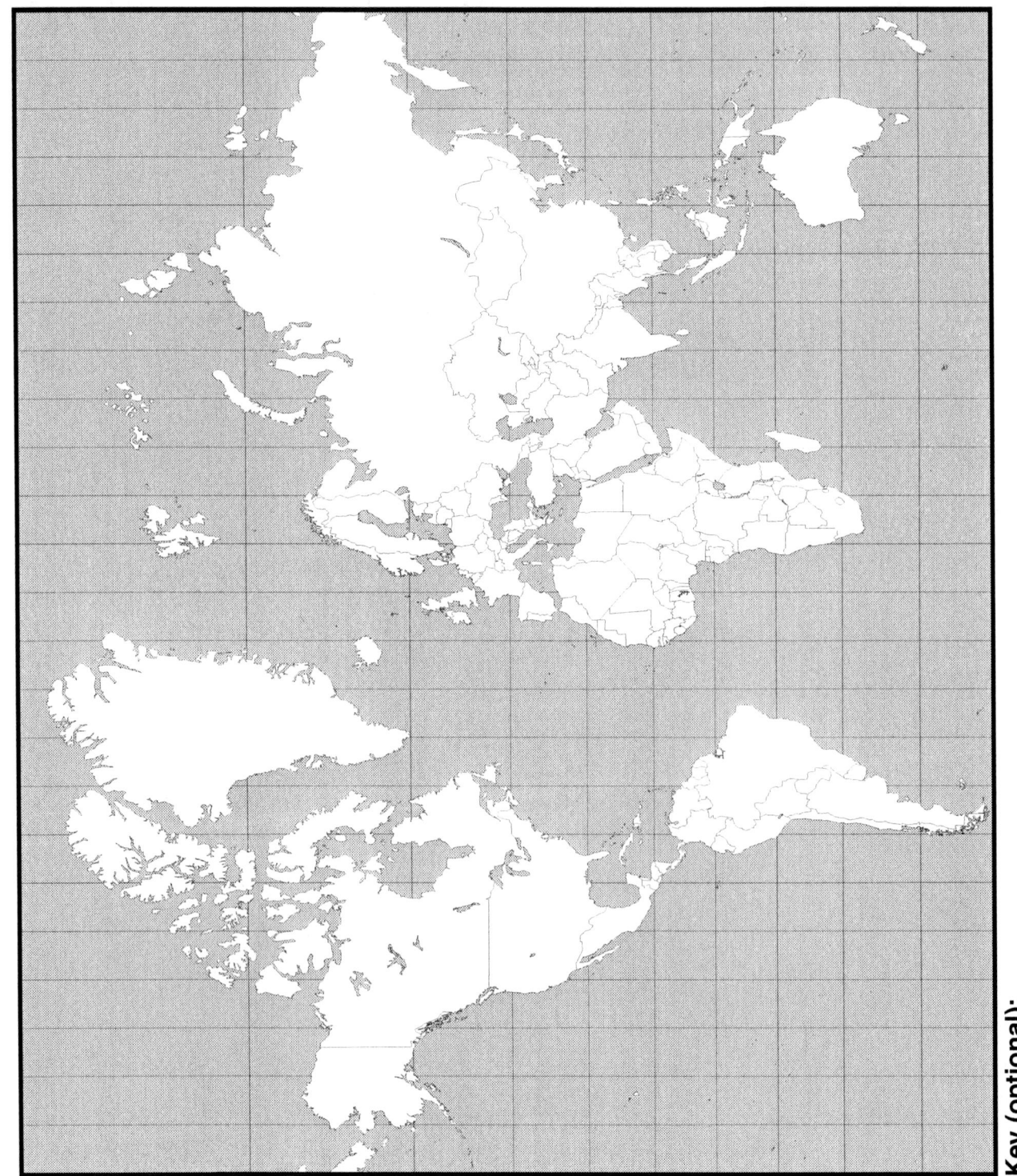

Key (optional):

Earth Science: Student Activity Page

Earthquake Patterns (cont.)

Directions: Refer to the earthquake location map you created to answer the first question below.

1. Look carefully at the earthquake locations you plotted. Describe any patterns you see in the locations of the earthquakes:

   _____
   _____
   _____
   _____
   _____
   _____
   _____
   _____

2. Now use the tectonic plates map below to describe the connection between the earthquake pattern(s) you discovered and plate tectonics. Write your description on the back of this page.

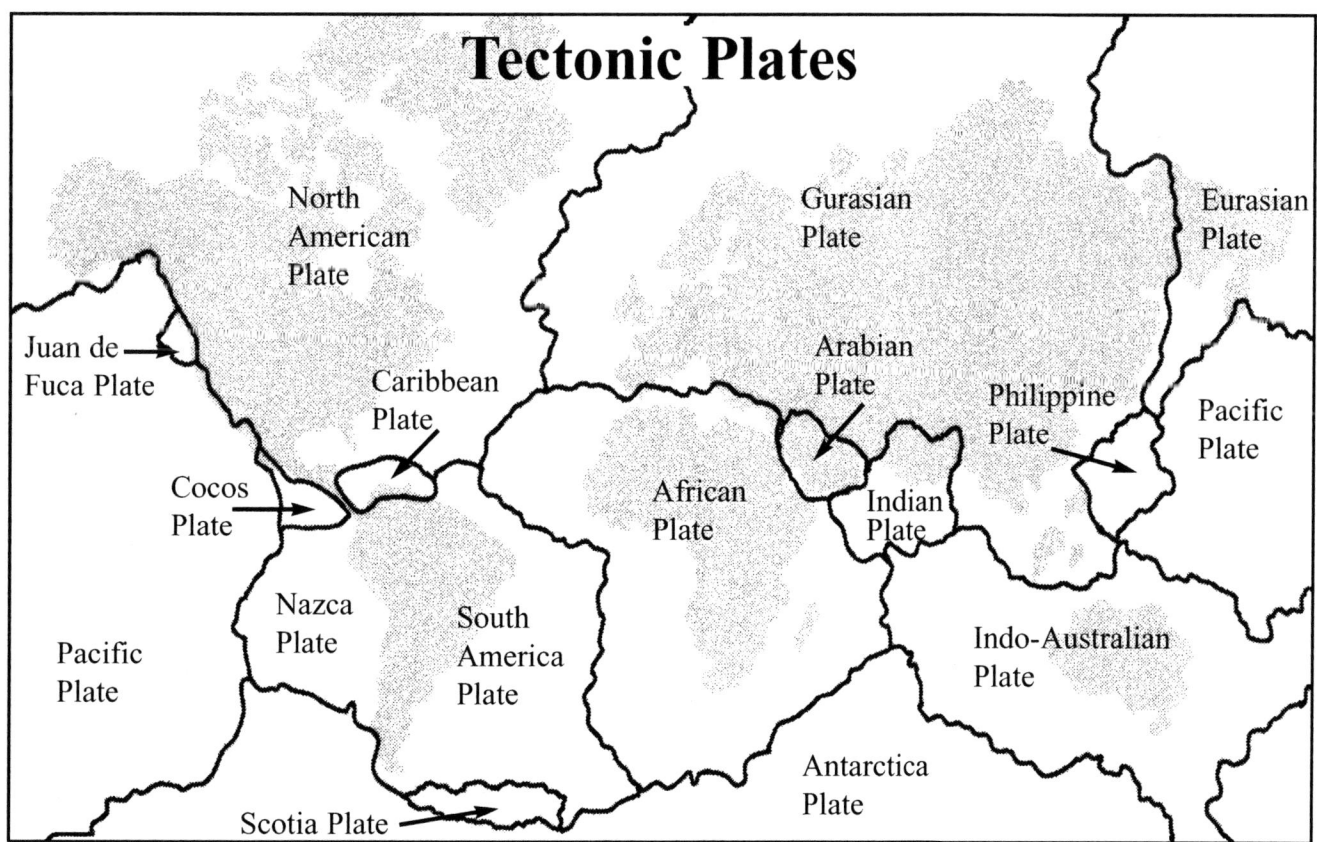

Earth Science: Teacher Notes

Saving the Environment***

NSTA Standards (5–8):
A, C, F

Objectives:
Students will:

- identify and research a current environmental problem
- record the effects of the problem on the environment/ecosystems and everyday life
- suggest possible solutions to the environmental problem

Suggested Web Site:
http://www.teachercreated.com/books/3842

Click page 26, site 1

Alternate Web Site(s):
Ask Jeeves for Kids

http://www.ajkids.com

Pre-Internet Activities:
- Brainstorm and discuss current environmental problems occurring across the globe on the first student activity sheet provided. As a class, identify the causes and effects of some of these problems.
- Write on the board the list of example environmental problems found below.
- When students have finished recording the causes and effects of an environmental problem on the second Saving the Environment student activity sheet, allow them to discuss with a partner some possible solutions to the problem. Then have them record at least two possible solutions in the space provided.

Extensions:
- Instruct students to find an organization on the Internet that is working toward solving the environmental problem they researched. What is the organization doing to solve the problem? Will the organization and others like it succeed in solving the problem? Why or why not?
- Direct students to research what governments across the world are doing to solve the environmental problem.

Example Environmental Problems:

Air pollution	Acid rain	Global warming	Ozone depletion
Water pollution	Hazardous waste	Rainforest depletion	Overpopulation

Earth Science: Student Activity Page

Saving the Environment *(cont.)*

Name: _____ Date: _____ Period: _____

Directions: Brainstorm environmental problems and their causes and effects using the chart below.

Environmental Problem	Possible Causes	Possible Effects

Earth Science: Student Activity Page

Saving the Environment *(cont.)*

Name: _____ Date: _____ Period: _____

Directions: Think of an environmental problem that concerns you or affects your daily life. If you live in a city, perhaps you are concerned about air pollution or overpopulation. Write the environmental problem that concerns you in the space below.

Go to http://www.teachercreated.com/books/3842
and click page 26, site 1. Using the Web site, find the cause(s) and effects of the environmental problem you wrote down. When you are finished, brainstorm possible solutions to the problem. Is there some small action you can take each day to help ease the problem?

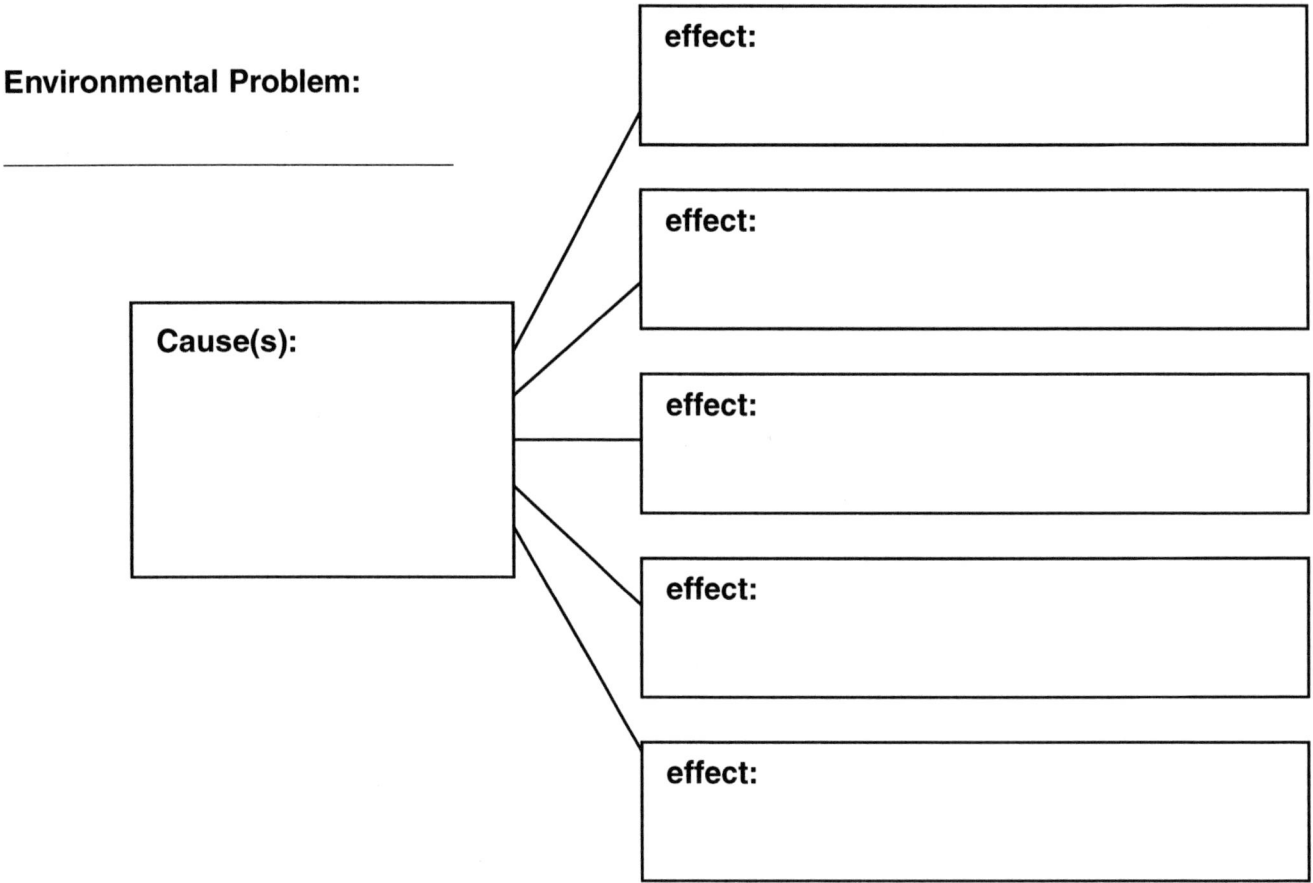

Possible solutions to this environmental problem:

Earth Science: Teacher Notes

Measuring Weather Conditions**

NSTA Standards (5–8):
A, D

Objectives:
Students will:
- chart the weather (temperature and precipitation) of four U.S. cities every day for one week
- compare weather in different parts of the country
- determine average temperatures and precipitation for each of the four cities
- hypothesize as to why one city has the least precipitation

Suggested Web Site:
http://www.teachercreated.com/books/3842

Click page 29, site 1

Alternate Web Site(s):
USA Today.com Weather Forecasts & News
http://asp.usatoday.com/weather/weatherfront.aspx

Pre-Internet Activities:
- Direct students to choose four major cities in the United States, locate and mark them on the map given, and enter the city names on the chart.
- Discuss the term *precipitation.*
- Check to be sure the suggested or alternate Web site is working. If both are not working, find another Web site by going to **http://www.google.com**, typing "weather" in the Search box, pressing Enter, and selecting a Web site from the list.
- Copy the two pages of the student activity sheet front to back, or staple them together, since the directions for both pages are spelled out on the first page.

Extensions:
- Create a class chart or graph of all the students' findings for average temperatures and precipitation. Determine the highest and lowest average temperatures and highest and lowest average precipitation for the entire class' data. Discuss each of these findings in small groups or as a class. For example, ask students why they think the city they found to have the lowest average temperature is generally colder than the rest of the cities. What affects the weather in that city?
- Discuss and compare the data for other cities while looking at a map of the country. Sample questions: How do the Great Lakes affect the weather in that part of the country? How do the temperature and precipitation in coastal cities compare with temperatures and precipitation in inland cities at the same latitude? How do temperatures compare at the same latitude on the West Coast and the East Coast? Even if students do not know the answers, it is an important exercise for them to hypothesize and attempt to back up their ideas with reasons.

Earth Science: Student Activity Page

Measuring Weather Conditions (cont.)

Name: _____ Date: _____ Period: _____

Directions: Find and mark four major cities in different parts of the country on the map below. Enter those city names on the chart on the next page. Next, go to
http://www.teachercreated.com/books/3842
and click page 29, site 1. Use the Web site to look up the daily temperature and precipitation for each city for five days. Note: for each city, record the temperature under "T" and the precipitation under "P" in the chart.

At the end of five days, find the average temperature and precipitation for each city and record it in the chart. In the space below the chart, note the cities with the highest average temperature and the highest average precipitation. Also note the cities with the lowest average temperature and lowest average precipitation. Then write an explanation for why you think the city with the lowest average precipitation has less precipitation than the other cities.

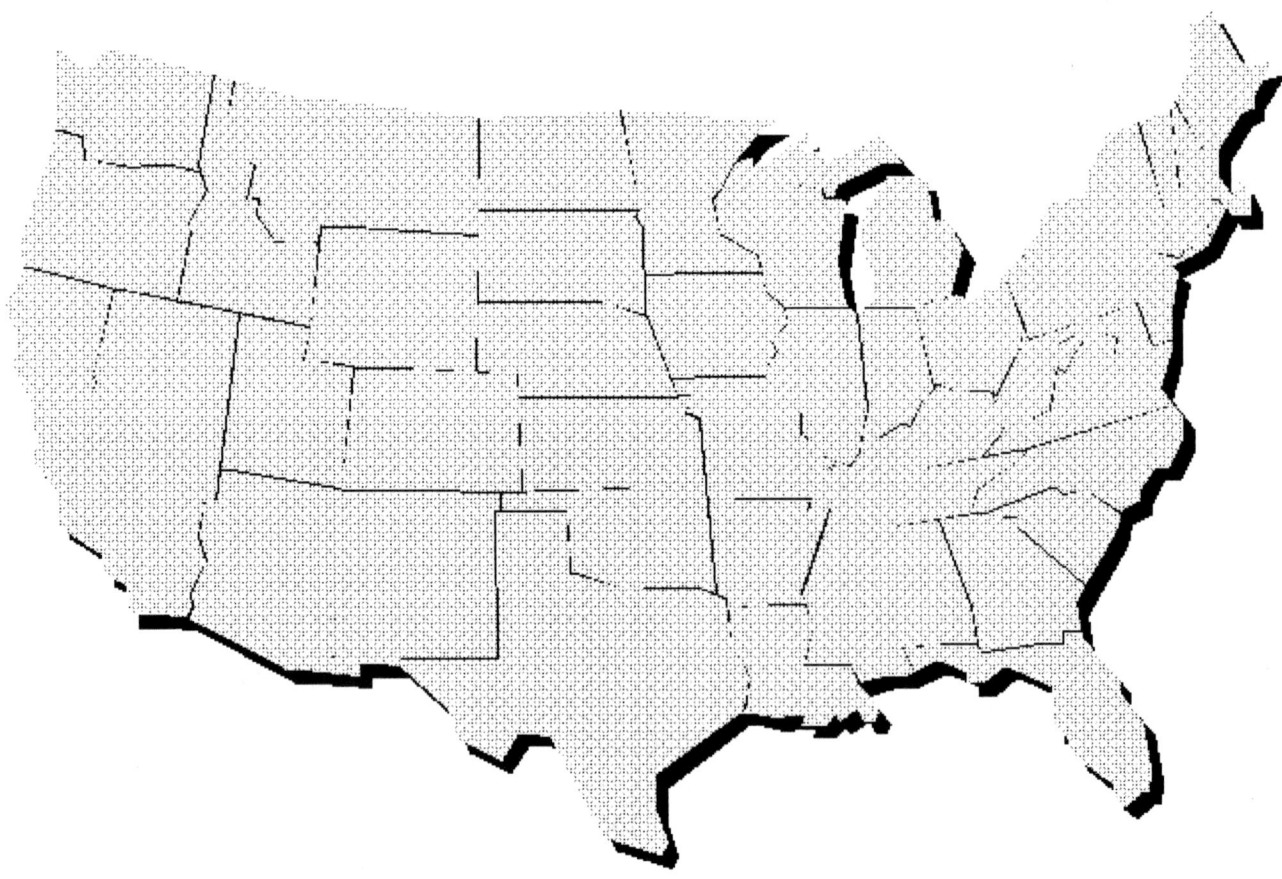

Earth Science: Student Activity Page

Measuring Weather Conditions (cont.)

Name: _____ Date: _____ Period: _____

Temperature and Precipitation Chart

City	Day 1		Day 2		Day 3		Day 4		Day 5		Average	
	T	P	T	P	T	P	T	P	T	P	T	P

Which city has the highest average temperature? _____

Which city has the lowest average temperature? _____

Which city has the highest average precipitation? _____

Which city has the lowest average precipitation? _____

Why do you think the city with the lowest average precipitation has less precipitation than the other cities?

_____

_____

_____

_____

Life Science: Teacher Notes

What's in a Name?**

NSTA Standards (5–8):
A, C, G

Objectives:
Students will:
- determine the common names for several familiar animals and plants given the scientific names
- understand the scientific naming system

Suggested Web Site:
http://www.teachercreated.com/books/3842
Click page 32, site 1

Alternate Web Site(s):
Yahoo
http://www.yahoo.com

AlltheWeb
http://www.alltheweb.com

Pre-Internet Activities:
- Brainstorm the reasons why scientists give animals and plants scientific names. Then have students read the text at the top of the What's in a Name student activity sheet to learn more about the scientific naming system.

Extensions:
- Have students choose one or two scientific names to investigate further. Direct them to an online or hard copy Latin dictionary to look up each of the scientific names. Then have them do a little research on each plant or animal to find out why scientists gave that organism its unique name.
- Have students come up with their own scientific names for common plants and animals. Stipulate that the names must somehow describe the animal or plant – its appearance, location, behavior, etc.

Life Science: Student Activity Page

What's in a Name? (cont.)

Name: _____ Date: _____ Period: _____

In 1758, Carl Linnaeus suggested that scientists use a universal naming system to identify plants and animals. In this way, scientists in different places could identify the plants and animals they studied by the same names, making it easier to discuss and share information.

Scientific names are made up of two Latin words, such as the name for the Polar Bear: Ursus maritimus. The first word is like a person's family name or last name. Ursus means "bear" in Latin, and is the scientific name for the family of bears. The second word is the particular species' name, and often describes the animal or plant in some way. For example, Maritimus is Latin for "of or on the sea," which is like the word maritime in English. Thus, an Ursus maritimus is a bear that is of the sea. Anyone who has been to a zoo knows that this is a good description of polar bears—they live a large portion of their lives swimming and hunting in the water.

Directions: Go to **http://www.teachercreated.com/books/3842**
and click page 32, site 1. Use this search engine to find the common names for these plants and animals. Type a scientific name in the Search box and then press Enter to bring up a list of sites that might tell you the common name. After you find each one, make a guess as to why the animal or plant was given its particular scientific name. Hint: Think about how the Latin words relate to English words that you know. The first one has been done for you.

Scientific Name	Common Name	Why was it given this scientific name?
Cereus giganteus	Saguaro cactus	I am guessing that Cereus giganteus is Latin for "gigantic cactus." I know Saguaro cacti can grow to be very tall, and the scientist who named it probably thought so too.
Acer macrophyllum		
Citrus limon		

©Teacher Created Materials, Inc.

Life Science: Student Activity Page

What's in a Name? *(cont.)*

Name: _____ Date: _____ Period: _____

Scientific Name	Common Name	Why was it given this scientific name?
Felis cattus		
Equus caballus		
Rhus toxicodendron		
Canis familiaris		

Bonus question: Why do you suppose Latin was chosen as the language for the scientific naming system?

Life Science: Teacher Notes

Life on Earth**

NSTA Standards (5–8):
A, C, D

Objectives:
Students will:
- find and record the names and dates of four fossils using the Internet
- draw a picture of each fossil found
- infer from the description of the fossil if the species is now extinct

Suggested Web Site:
http://www.teachercreated.com/books/3842
Click page 35, site 1

Alternate Web Site(s):
Enchanted Learning Finding Dinosaur Fossils
http://www.zoomschool.com/subjects/dinosaurs/dinofossils/Fossilfind.html

Pre-Internet Activities:
- Discuss how fossils tell us that many species of animals and plants once lived long ago but do not exist today. Species become extinct when the environment changes and the animal or plant cannot adapt to the new conditions. However, some fossils tell us that certain species of plants and animals that lived long ago still exist today. These organisms were able to adapt as the environment changed around them.
- Briefly discuss how to determine if a fossil species is extinct, since it may not explicitly state this in the fossil's description. In some cases, students will have to infer as to whether the species survives today. For example the description says something like, "Ammonites *lived* in oceans near the land masses..." instead of "Ammonites *live* in oceans near the land masses..." then most likely the species is extinct.

Extensions:
- Using the Internet, determine the best places to find fossils. What types of rocks are fossils found in, and where is that rock accessible to fossil hunters? Why is erosion sometimes a fossil hunter's best friend? What states have exceptionally rich fossil histories? If possible, find a local place to go fossil hunting.
- Discuss how fossils are formed and what can be fossilized. What are some of the most common fossils and what are some of the rarest?

Life Science: Student Activity Page

Life on Earth (cont.)

Name: _____ Date: _____ Period: _____

Directions: Go to **http://www.teachercreated.com/books/3842**
and click page 35, site 1. Find the names and draw the pictures of four fossils in the spaces provided. Finally, determine if each species is extinct.

Fossil name: _____

Period/Epoch/Date: _____

Is species extinct (Y/N)? _____

Picture:

Fossil name: _____

Period/Epoch/Date: _____

Is species extinct (Y/N)? _____

Picture:

Fossil name: _____

Period/Epoch/Date: _____

Is species extinct (Y/N)? _____

Picture:

Fossil name: _____

Period/Epoch/Date: _____

Is species extinct (Y/N)? _____

Picture:

Life Science: Teacher Notes

Trees are More than Wood**

NSTA Standards (5–8):
A, C, F

Objectives:
Students will:
- examine the shapes of trees and find examples of different shapes on the Internet
- hypothesize as to why trees have different shapes
- understand the relationship between tree shape and survival
- evaluate their original hypotheses about tree shape

Suggested Web Site:
http://www.teachercreated.com/books/3842
Click page 37, site 1

Alternate Web Site(s):
All the Web
http://www.alltheweb.com (Hint: Click the **Picture** tab before searching.)

Pre-Internet Activities:
- As a class, brainstorm and list different types of trees and ask students to either draw each one on a sheet of paper or picture each one in their minds.
- Explain to students that they will be using the image finder feature of a search engine. The search engine will return images as the results, not a list of Web sites.
- Review how to form a hypothesis.

Extensions:
- Before explaining why trees have different shapes, ask students to find five examples of trees that grow far north and five examples of trees that grow near the equator. Ask them to draw each group of five and then make observations about the trees, including tree shape. Next, ask them to find similarities and differences, and then revise their original hypotheses about tree shape if necessary.

Explanation as to Why Trees have Different Shapes:
Most trees grow in a shape that maximizes the amount of sunlight their leaves collect. Trees near the equator, where the noontime sun is directly overhead, tend to have flat tops that expose the maximum number of leaves to the sun and thus collect the maximum amount of light. Trees farther north, closer to the Arctic Circle, tend to be cone-shaped because the sun is often low in the sky. The leaves on these trees cover the tree from top to bottom, thus exposing the tree to the sun that is on the "side" of the tree. Finally, understory trees (those that are shadowed by taller trees) are often round-shaped because light filtered by taller trees comes from many different directions. And of course, trees that are partially blocked from sunlight will have more leaves (and an asymmetrical shape) on the side exposed to the sun. (Paraphrased from the Web site, "What Shapes Trees?" at
http://www.units.muohio.edu/dragonfly/itd/learn.htmlx)

Life Science: Student Activity Page

Trees are More than Wood *(cont.)*

Name: _____ Date: _____ Period: _____

Directions: Follow the steps below.

1. Think about a tree, any tree. Draw the shape of this tree in the space below.

2. Are all trees shaped this way? Explain your answer.
   _____
   _____

3. Go to **http://www.teachercreated.com/books/3842**
   and click page 37, site 1. Use this image search engine to find two trees that are shaped differently from the tree you drew above. Draw and name those trees in the space provided below.

   Name of tree:

   Name of tree:

Life Science: Student Activity Page

Trees are More than Wood (cont.)

Name: _____ Date: _____ Period: _____

4. Brainstorm a list of reasons why trees grow in different shapes and write those reasons here:

   _____
   _____
   _____
   _____
   _____

5. Now form a hypothesis that explains why trees grow in different shapes and write that here (in complete sentences).

   _____
   _____
   _____
   _____
   _____

6. Your teacher will read the explanation of why trees grow in different shapes. After you have heard the explanation, describe how your hypothesis should be changed:

   _____
   _____
   _____
   _____
   _____

Life Science: Teacher Notes

Your Local Endangered Species**

NSTA Standards (5–8):
A, C, F

Objectives:
Students will:
- examine the idea that extinction happens when a species cannot adapt to changing conditions in its environment
- use the Internet to find three organisms in their state which are endangered
- conduct Internet research to find out more about one endangered species and its current status
- synthesize information found on an endangered species to make a case for saving it
- develop a plan for saving the endangered species

Suggested Web Site:
http://www.teachercreated.com/books/3842
Click page 40, site 1

Alternate Web Site(s):
Threatened/Endangered Species Map
http://www.indiana.edu/~bradwood/eagles/threatened_endangered.htm

Pre-Internet Activities:
- Discuss the concept of *extinction*. Explain that extinction happens when a species cannot adapt to changing conditions in its environment. Sometimes extinction is brought about by natural disasters or naturally occurring changes. Other times human action threatens the survival of a species. Brainstorm the human actions that can threaten the survival of a species (ideas: destruction of habitat, pollution of water or food sources, global warming, overpopulation, etc.).
- Discuss the difference between a threatened species and an endangered species. *Threatened* species face serious challenges to their survival, but are not yet in danger of extinction. An *endangered* species is on the brink of extinction. Explain that today the class is focusing on endangered species.
- Briefly discuss in general terms what actions humans have taken to save endangered species, such as land conservation, imposing regulations on industry to control pollution, creating the threatened and endangered species lists, educating the public, regulating hunting and fishing, etc.

Extensions:
- In small groups, direct students to share the information they collected on their endangered species. Then have them discuss their opinions on why the animals and plants they chose are important to save. You might lead students in a discussion of how entire ecosystems can be affected by the extinction of one organism. Illustrate this with a drawing of a food web.

Life Science: Student Activity Page

Your Local Endangered Species (cont.)

Name: _____ Date: _____ Period: _____

1. Go to **http://www.teachercreated.com/books/3842**
   and click page 40, site 1. Use the Web site to find three endangered species (plants or animals) in your state. To do this, type the keywords "endangered species" and then the name of your state. For example, if you live in California, type "endangered species California" in the search box. View Web sites from the results list until you have found three endangered species in your state, and write their names here:

   _____   _____   _____

2. Now choose one of the endangered species you found above and investigate it further. Write its name in the first row of the table below. Then using the same search engine, type the animal or plant's name in the Search box and press Enter. From the list that appears, choose Web sites that will help you find and record the facts requested in the table below. If you cannot find some of the facts, record other interesting facts you find about your plant or animal next to "Other facts:."

Name of the endangered plant or animal:
Current number in the wild:
What it eats:
Habitat (where it lives):
Why it is endangered:
What people are doing to save this endangered species:
Other facts:

Life Science: Student Activity Page

Your Local Endangered Species (cont.)

Name: _____ Date: _____ Period: _____

3. Write three reasons why it is important to save this endangered species. (If you cannot find three reasons, think of possible reasons on your own.)

   _____
   _____
   _____
   _____

4. Imagine that your community has asked you to think of a way of saving the endangered species you researched. What would you do? Think of something new or pick the best idea from your research. Create your plan below:

The endangered species I plan to save:
What I would do to save it:
I would need help from these people:

Life Science: Teacher Notes

Classifying Living Things**

NSTA Standards (5–8):
A, C, F

Objectives:
Students will:
- gather data on frogs using the Internet
- analyze the frog data to find similarities and ways of grouping
- develop a classification system for frogs based on several characteristics

Suggested Web Site:
http://www.teachercreated.com/books/3842
Click page 43, site 1

Alternate Web Site(s):
Exploratorium: Frogs
http://www.exploratorium.edu/frogs/

Pre-Internet Activities:
- Gather animal and plant identification books to show to the students.
- Explain that scientists group animals and plants according to similarities. This is called *classification*, and you can see how scientists do this by looking at an identification book. For example, you might look at a bird identification book and note that the birds are grouped by similar colors. Or perhaps certain birds are grouped together because they sing (e.g., songbirds). Or maybe there is a group of birds that all live in the jungle. Common ways that scientists group animals are by color, where they live (environment), sounds they make, size, food they eat, or how they defend themselves (poison, camouflage, fangs, claws, etc.).

Extensions:
- Ask students the following question. "Why do scientists develop classification systems?" Have students discuss in small groups and share with the rest of the class, or discuss the question as a whole class.
- Obtain several identification books that can be used to identify organisms (birds, regional wildlife, flowering plants, insects, etc.) in your area. Put students in small groups, give them clipboards and paper to write on, and head outside. Let students hunt for specimens. Instruct them that once they find an animal, plant, insect, etc. that they should look it up in their books. Have them record the name and a few characteristics of each specimen. When they are finished, ask students to share what they found with other groups or the whole class.

Life Science: Student Activity Page

Classifying Living Things (cont.)

Name: _____ Date: _____ Period: _____

Directions: Go to **http://www.teachercreated.com/books/3842** and click page 43, site 1. Use the subject directory to research frogs. View more than one Web site if necessary to find information on several frogs. For each frog you find, write its name down in the chart below. Then write down its color, where it lives (in the Environment column), what sound(s) it makes, its size (in inches, if possible) and the way it defends itself (in the Defense column). If your teacher gives you permission, print and save a picture of each frog you find. **Hint:** Frogs are amphibians.

Name	Color	Environment	Sound(s)	Size (inches)	Defense
Frog 1:					
Frog 2:					
Frog 3:					
Frog 4:					
Frog 5:					
Frog 6:					
Frog 7:					
Frog 8:					
Frog 9:					
Frog 10:					

Life Science: Student Activity Page

Classifying Living Things *(cont.)*

Name: _____ Date: _____ Period: _____

Directions: Look at the information you found on frogs in the chart on the previous page. Look for similarities among the frogs and create at least three groups of frogs based on those similarities. For example, you might choose to create a group of all green frogs, a group of frogs that make the same sound, and a group of frogs that are poisonous. It's all right if some frogs are in more than one group. Be sure to give each group of frogs a name.

Group 1

Name of Group: _____

Frogs in this group:

Group 2

Name of Group: _____

Frogs in this group:

Group 3

Name of Group: _____

Frogs in this group:

Life Science: Teacher Notes

Wildflowers in Your State*

NSTA Standards (5–8):
A, C

Objectives:
Students will:
- find, identify, draw, and describe four wildflower species in the students' state

Suggested Web Site:
http://www.teachercreated.com/books/3842
Click page 46, site 1

Alternate Web Site(s):
Yahoo
http://www.yahoo.com

Pre-Internet Activities:
- Ask students if they have ever noticed wildflowers growing along roads, in their yards, at the park, on the playground, or even in the ruts of the sidewalk. Some of these plants look like weeds, but actually, many are native wildflower species. It can be fun to try to find and identify wildflowers – it's like finding a buried treasure! Tell students that today they will find wildflowers that grow in their state, but they won't be looking outside. They'll be using the Internet.

Extensions:
- After the students have viewed and drawn several wildflowers, allow them to try to find them outside. The school grounds are a good place to look for wildflower species. First, create groups of 3-4 students. Then direct the students to share their drawings and descriptions of different wildflowers. Inform them that they can look for all of the wildflowers in their group. Then tell the students where they can go to look for wildflowers, and ask them to keep a record of any that they find. The record can be as simple as making a tally mark next to each wildflower picture, or students can collect and preserve the specimens. If students collect the specimens, give them plastic sealable bags to carry them in. Then when they are back in the classroom, have them press the wildflowers using heavy books and paper towels. When pressed, create a display using poster board of all of the wildflowers (labeling each one with its common name and scientific name) and hang it on a bulletin board or keep it at your science center.
- For homework, ask students to collect five flower specimens of any kind. Then, using flower identification books or the Internet, challenge students to identify all of the flowers (this can also be done at home using library books or the Internet) and record their origins. Then have students share their flowers and the origins of the flowers, noting which ones are native and which ones are non-native. Most likely, many of the flowers will be non-native. Ask students to discuss why many of the flowers are non-native. How did the non-native flowers get here? What effects, if any, does the presence of non-native flowers have on native species?

Life Science: Student Activity Page

Wildflowers in Your State *(cont.)*

Name: _____ Date: _____ Period: _____

Directions: Go to **http://www.teachercreated.com/books/3842** and click page 46, site 1. Use the Web site to find wildflowers that grow in your state. Try using the keywords "wildflowers" and the name of your state. For example, if you live in Texas, you would type "wildflowers Texas" in the search engine Search box. Once you have found a Web site that shows pictures of wildflowers in your state, record the common and scientific names of the flower, and write a description of the flower based on how it looks. Then draw a picture of the flower.

Wildflower #1

Common Name:
_____

Scientific Name:
_____

Description:
_____
_____
_____
_____

Picture:

Wildflower #2

Common Name:
_____

Scientific Name:
_____

Description:
_____
_____
_____
_____

Picture:

Life Science: Student Activity Page

Wildflowers in Your State *(cont.)*

Name: _____ Date: _____ Period: _____

Wildflower #3

Common Name:

_____

Scientific Name:

_____

Description:

_____

_____

_____

Picture:

Wildflower #4

Common Name:

_____

Scientific Name:

_____

Description:

_____

_____

_____

Picture:

Which one of the four flowers is your favorite? Why?

Life Science: Teacher Notes

Ecosystem Research**

NSTA Standards (5–8):
A, C

Objectives:
Students will:
- research a particular ecosystem
- identify the ecosystem's producers, consumers, and decomposers
- show in a food web diagram how the producers, consumers, and decomposers are interdependent

Suggested Web Site:
http://www.teachercreated.com/books/3842
Click page 49, site 1

Alternate Web Site(s):
The Desert Ecosytem
http://www.windows.ucar.edu/tour/link=/earth/desert_eco.html&edu=elem

Internet Geography – Ecosystems
http://www.bennett.karoo.net/topics/ecosystem.html

Pre-Internet Activities:
- Brainstorm and discuss in groups or as a class each of the following concepts: ecosystem, producer, consumer, and decomposer. Explain the roles of producers (make their own food and provide food for consumers and decomposers), consumers (eat other organisms), and decomposers (eat dead plants and animals, and put nutrients back into the soil that producers use to make their own food) in an ecosystem. Give students some examples of each concept (examples are listed below), and to show how producers, consumers, and decomposers are interdependent in an ecosystem. (Use a food web diagram like the one on page 52.)
Then individually or in groups, allow students to try to form definitions of each concept using the student activity sheet on the following page. (Note: you may choose to allow students to use the Internet to do the following activity sheet.)

Ecosystems:	Producers:	Consumers:	Decomposers:
desert	milkweed plant	eagle	bacteria
ocean	oak tree	fox	fungus
savannah	grass	snake	
rainforest	cactus	insects	

Extensions:
- Creating a miniature ecosystem (a terrarium) is an excellent way for students to learn in a hands-on way how an ecosystem functions. Search online using the keyword "terrarium" to find directions on how to build a terrarium out of simple materials.

Life Science: Student Activity Page

Ecosystem Research (cont.)

Name: _____ Date: _____ Period: _____

Directions: After your class has brainstormed and discussed ecosystems, producers, consumers, and decomposers, answer the following questions.

1. What is an ecosystem? _____
   _____
   _____
   _____

2. What is a producer? _____
   _____
   _____
   _____

3. What is a consumer? _____
   _____
   _____
   _____

4. What is a decomposer? _____
   _____
   _____
   _____

5. Explain why producers, consumers, and decomposers are dependent upon one another in an ecosystem: _____
   _____
   _____
   _____

Life Science: Student Activity Page

Ecosystem Research (cont.)

Name: _____ Date: _____ Period: _____

Directions: Choose a type of ecosystem to research, go to
http://www.teachercreated.com/books/3842
and click page 49, site 1. Use the Web site to complete this activity page. To find information on your ecosystem, type the name of the ecosystem followed by the word "ecosystem" in the Search box (for example, to find information on the desert ecosystem, type "desert ecosystem" in the Search box) and press Enter. Choose the best Web site from the list. Note: you may need to look at more than one Web site to find all of the information.

1. Circle the type of ecosystem you will research:

   desert   ocean   estuary   lake   coniferous forest

   grasslands   tundra   wetlands   rainforest   deciduous forest

   river   savannah

2. Name some of the plants (producers) that grow in this ecosystem: _____

   _____

   _____

   _____

3. Name some of the insects and animals (consumers) that live in this ecosystem:

   _____

   _____

   _____

4. If possible, try to find the names of the decomposers that live in this ecosystem. If you cannot find the names, make a guess as to some of the decomposers that can be found there.

   _____

   _____

   _____

Life Science: Student Activity Page

Ecosystem Research (cont.)

Name: _____ Date: _____ Period: _____

Directions: Enter some of the plants, insects, animals, and decomposers you listed on the previous page in the appropriate circles below. Be sure to show true food relationships between the organisms.

Type of Ecosystem: _____

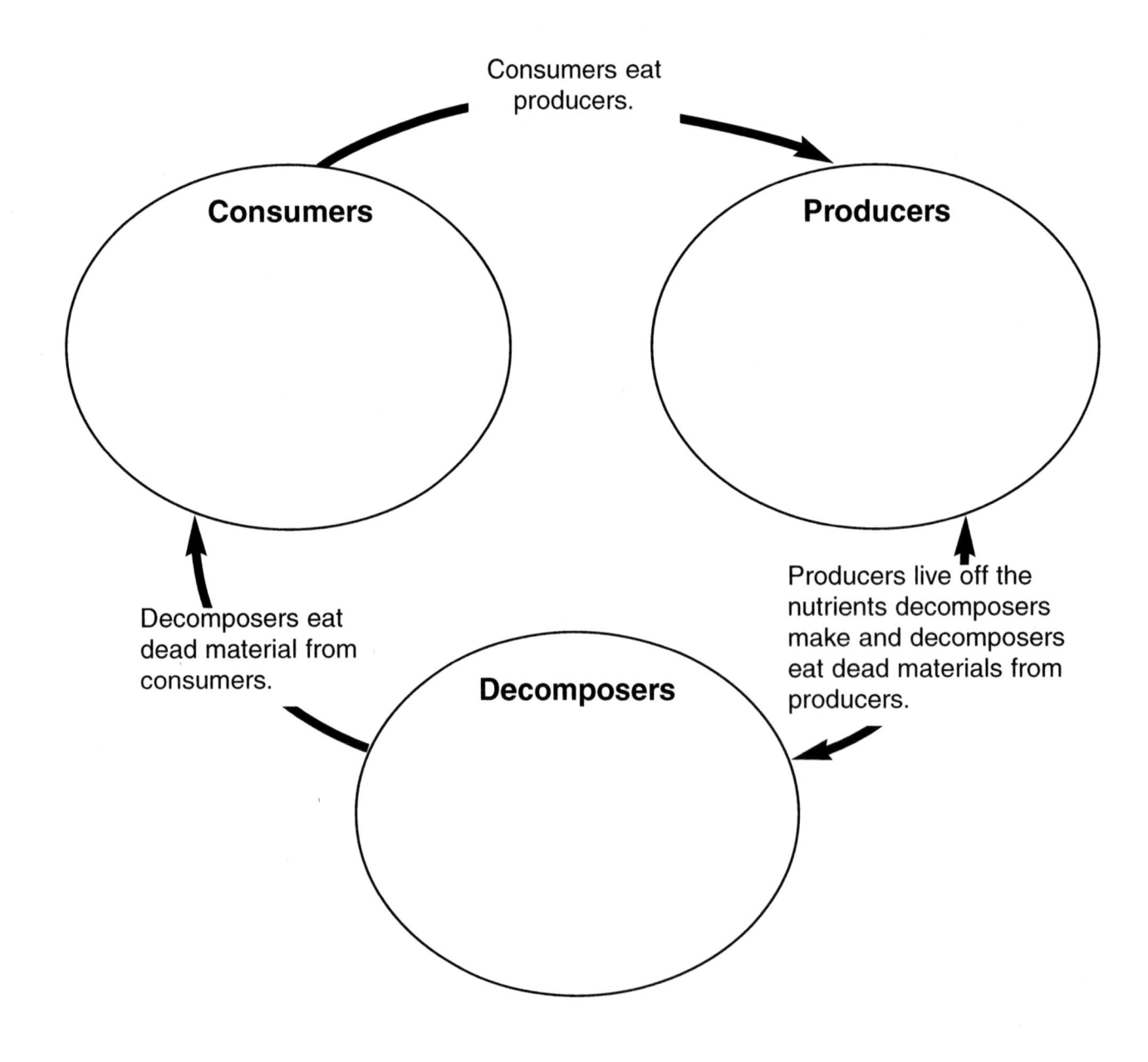

Consumers eat producers.

Consumers

Producers

Decomposers eat dead material from consumers.

Decomposers

Producers live off the nutrients decomposers make and decomposers eat dead materials from producers.

Life Science: Teacher Notes

Myrmecology: The Study of Ants**

NSTA Standards (5–8):
A, C

Objectives:
Students will:
- find a drawing of an ant's anatomy on the Internet and label the major parts of an ant
- name at least 3 different species of ants
- research one species of ant in more depth

Suggested Web Site:
http://www.teachercreated.com/books/3842

Click page 53, site 1

Alternate Web Site(s):
Google

http://www.google.com

Pre-Internet Activities:
- Ask students if they have ever observed or read about ants. What were they doing? What did they look like? Point out that there are many different species of ants in the world. Each species is unique in its behavior, where it lives, what it eats, etc.

Extensions:
- Allow students to venture outside to observe ants in action. Provide them with clipboards and paper to write down their observations of ants (and if possible, also provide them with rulers to measure one ant. Magnifying glasses are also handy.). Consider having them copy down the following questions before they go to focus their observations:
  - Follow the ants' trail. Where are they coming from and where are they going? Where do you think their nest or colony is located?
  - Are the ants collecting or carrying anything? If so, what?
  - Describe the ants' appearance, including color, size, and any distinctive features. If you have a ruler, try to measure one ant. (Hint: You might have to find a dead ant in order to take a measurement.)
  - Do you see any ants that appear different than the others? If so, describe how they look different.

  After students have collected their data, instruct them to use the Internet to try to identify the ant species they found. Suggest that they use a search engine and type "ant database" or "ant identification" into the Search box to find a Web site that might help them.

Life Science: Student Activity Page

Myrmecology: The Study of Ants *(cont.)*

Name: _____ Date: _____ Period: _____

Directions: Go to **http://www.teachercreated.com/books/3842** and click page 53, site 1. Use the search engine to find an illustration of an ant's anatomy. To do this, type "ant anatomy" in the search engine's Search box. Choose the best Web site from the list. Then label the parts of the ant shown below.

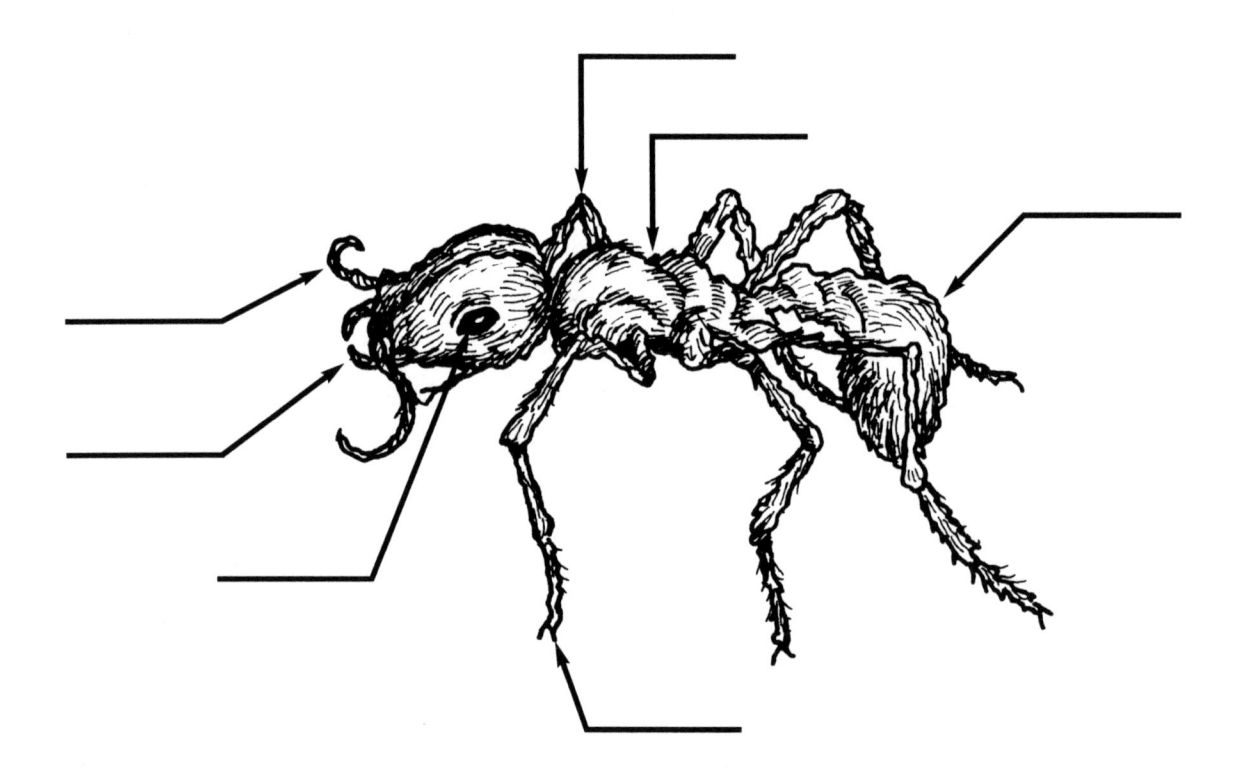

TCM# 3842 Web Resources for Science Activities   ©*Teacher Created Materials, Inc.*

Life Science: Student Activity Page

Myrmecology: The Study of Ants *(cont.)*

Name: _____ Date: _____ Period: _____

Directions: Use the subject directory on the same Web site (go to
http://www.teachercreated.com/books/3842
and click page 53, site 1) to find Web sites that will help you answer the questions about ants below.

Name three species of ants:

Read more about one species of ant and answer the following questions:

What does this species of ant eat?
Describe the colony or nest of this species of ant:
What are the roles (queen, workers, etc.) of the ants of this species?
Indicate one interesting fact about this species of ant:
Indicate another interesting fact about this species of ant:

Life Science: Teacher Notes

Insects*

NSTA Standards (5–8):
A, C

Objectives:
Students will:
- observe insects and record their observations
- share the types of insects they found and their observations with students across the world

Suggested Web Site:
http://www.teachercreated.com/books/3842

Click page 56, site 1

(Use the suggested Web site to find e-pals for your students and do collaborative projects with classrooms across the world.)

Alternate Web Sites:
KeyPals Club at Teaching.com

http://www.teaching.com (Another site to help you find e-pals for your students)

Pre-Internet Activities:
- Find e-mail pen pals for your students before beginning this activity. Ideally, you will find another class that will share their observations of insects with your students. Before you begin corresponding about insects, it's a good idea to allow students to write introductory letters.
- Allow students to observe insects at home, around the school, and/or bring them to class in containers. For each insect, students should record their observations on the first Insects student activity sheet. If possible, direct students to identify the insect species name using identification books or the Internet.
- After students have recorded their observations of at least two insects, ask them to compose a letter on the second Insects student activity sheet. This will be the e-mail they will send to their e-pals.
- Note that this activity can be on-going. After students have written their e-pals and the e-pals have written back, have them do it again for as long as your study of insects continues.

Extensions:
- Create a data sheet that students can use to record information about the insects their e-pals share with them (or have them use the one provided). Challenge students to try to find pictures of the insects on the Internet based on the information their e-pals shared (insect species name is particularly helpful!). Eventually direct students to create a book of the insects they identified as well as the insects their e-pals shared with them.
- When answers are received from the other schools, post the letters on the bulletin board so students can read and learn about additional insects. After you have corresponded for a while, ask students to draw some conclusions about insects in other parts of the world. How are the insects similar or different?

Life Science: Student Activity Page

Insects *(cont.)*

Name: _____ Date: _____ Period: _____

Directions: Record each insect you find on the data sheet below. Be sure to draw it and then describe what it looks like, note where you found it, and record what it was doing when you saw it.

Insect #_____	Insect #_____
Name:	**Name:**
Drawing: **What it looks like (in words):**	**Drawing:** **What it looks like (in words):**
Where I found it (under a bush, on a leaf, etc.):	**Where I found it (under a bush, on a leaf, etc.):**
Behavior (what it was doing):	**Behavior (what it was doing):**

©Teacher Created Materials, Inc.

Life Science: Student Activity Page

Insects *(cont.)*

Name: _____ Date: _____ Period: _____

Directions: Use the form below to write a draft of an e-mail that you will send to your e-pal. An example letter is shown.

Example Letter

> Dear Susan,
>
>   Hi! How are you? I am writing to you about some insects I saw recently.
>   When I was walking to school the other day, I saw a praying mantis. It was long and stick-like. Its eyes seemed too big for its body. Its front legs were folded so that it looked like it was praying, and it was resting on a leaf. I can't tell you what species of praying mantis it was because I couldn't find it on the Internet.
>   The other insect I saw this week was a flea. It must have been a dog flea because it was crawling on my dog Sammy. The flea was very small and black. When I looked at it closely, I could see it had several tiny legs.
>   I hope you think these insects are interesting. Write back soon and tell me about your insects!
>
> Sincerely,
>
> Amy

Your Letter

> Dear _____,
>
>
>
>
>
>
>
> Sincerely,
>
> _____
> (Your name)

Life Science: Teacher Notes

Vertebrates & Invertebrates*

NSTA Standards (5–8):
A, C

Objectives:
Students will:
- find examples of vertebrates and invertebrates on the Internet
- find, draw, and label four different species of butterflies
- compare the unique characteristics of the vertebrate groups using a comparison matrix

Suggested Web Site:
http://www.teachercreated.com/books/3842
Click page 59, site 1

Alternate Web Site(s):
KidsClick!
http://sunsite.berkeley.edu/KidsClick!/

Pre-Internet Activities:
- Explain to students the difference between a *vertebrate* and an *invertebrate* animal. Vertebrates have backbones or spines, and invertebrates do not. Name a few vertebrates, such as humans, sharks, frogs, and eagles, and a few invertebrates, such as clams, spiders, sea anemones, and scorpions.
- Before beginning the second activity, challenge students to name the different groups of vertebrate animals. These groups are mammals, reptiles, amphibians, fish, and birds. Point out that these animals are separated into these groups because of major differences among them. It will be the students' job to find these differences on the Internet and complete the student activity page.

Extensions:
- Challenge students to draw a tree of the animal kingdom, showing how the first major classification is vertebrates and invertebrates, and then how each of these groups is divided. The interesting part of this activity is exploring how the invertebrates are divided, since students are already familiar with the vertebrate groups of mammals, reptiles, amphibians, fish, and birds.
- Direct students to explore the world of arthropods. Arthropods make up 90% of the animal kingdom, so they are worth studying! Have students determine what makes an arthropod an arthropod, such as an exoskeleton and jointed legs. Also direct them to find how arthropods are divided into subgroups (insects, arachnids, crustaceans, etc.). Discuss how it is possible that creatures such as lobsters and ants are part of the same major group. Then challenge them to think of invertebrate animals that are not arthropods. Finally, brainstorm the reasons why this major group of animals is so successful on earth, making up 90% of the animal kingdom.

Life Science: Student Activity Page

Vertebrates & Invertebrates (cont.)

Name: _____ Date: _____ Period: _____

Directions: Go to **http://www.teachercreated.com/books/3842** and click page 59, site 1. Use the Web site to find and create a list of vertebrates and invertebrates. To do this, first type "vertebrates" in the search engine Search box. Choose Web sites that list the names of vertebrates and then write those names in the Vertebrates column. Then type "invertebrates" in the search engine Search box to find Web sites that have the names of invertebrates. Write those names in the Invertebrates column.

Vertebrates	Invertebrates

Life Science: Student Activity Page

Vertebrates & Invertebrates *(cont.)*

Name: _____ Date: _____ Period: _____

Directions: Use the subject directory on the same Web site (go to
http://www.teachercreated.com/books/3842
and click page 59, site 1) to find examples of four different butterflies. Draw and label one butterfly in each of the boxes below. Hint: to find a list of butterfly Web sites in the subject directory, try clicking Animals, then Insects.

Butterfly name: _____

Butterfly name: _____

Butterfly name: _____

Butterfly name: _____

Bonus question: Are butterflies vertebrates or invertebrates? _____

©Teacher Created Materials, Inc.

Life Science: Student Activity Page

Vertebrates & Invertebrates (cont.)

Name: _____ Date: _____ Period: _____

Directions: Use the subject directory on the same Web site (go to **http://www.teachercreated.com/books/3842** and click page 59, site 1) to complete the matrix below. For each group of vertebrates, find a Web site in the subject directory that will help you determine its characteristics. Then place a check mark next to the characteristics that apply to each group of vertebrates. (**Hint:** Click **Animals** or **Living Things** in the subject directory to get started.)

Comparing Vertebrates

	Warm-blooded	Cold-blooded	Live birth	Lay eggs	Lungs	Gills	Rough, scaly skin	Wings and feathers	Fins and scales	Do not care for young	Nurse young and/or care for young
Mammals											
Reptiles											
Amphibians											
Fish											
Birds											

Use the chart to answer the following questions:

1. Which vertebrate groups are cold-blooded?

2. Which vertebrate groups nurse and/or care for their young?

Life Science: Teacher Notes

Animal Cam***

NSTA Standards (5–8):
A, C, E, F

Objectives:
Students will:
- observe and record animal behavior over a three-day period using an animal cam on the Internet
- synthesize the data to describe surprising animal behaviors and what was learned
- make an inference as to why scientists make observations of animals

Suggested Web Site:
http://www.teachercreated.com/books/3842
Click page 63, site 1

Alternate Web Site(s):
Discovery Channel Cams
http://dsc.discovery.com/cams/cams.html

San Diego Zoo
http://www.sandiegozoo.org/

Pre-Internet Activities:
- Locate an appropriate animal cam for your students before the activity begins. Type "animal cam" in the search engine Search box if you are looking for any animal to observe. View the animal cams on several Web sites from the list to choose the best one. Or, type the animal's name and then "cam" to find a specific animal (or insect). For example, if you are looking for a panda cam, type "panda cam" in the search engine Search box. Write the Web site address on the board for students.
- Note that there will be better times to observe certain animals. Many animals are more active early in the morning or toward dusk (and some are even nocturnal). Your animal choice may depend upon the time of day you are able to do this activity with your students. Obviously, active animals will be more interesting to the students.
- Discuss how to make scientific observations with your students. Tell students that good scientists make observations with all of their senses: sight, hearing, touch, smell, and even taste, if appropriate. However, sometimes scientists can't use all of their senses, so they rely on just a few. Today students will be able to use the senses of sight and possibly hearing, since they will be looking at a live video of an animal.

Extensions:
- Challenge students to research how the animal they observed is cared for. Direct them to research what the animal is fed, when it is fed, and what is provided to the animal to facilitate playing, sleeping, eating, resting, mating, etc. If the animal is potentially dangerous, have the students investigate how its keepers care for it without getting hurt.

Life Science: Student Activity Page

Animal Cam (cont.)

Name: _____ Date: _____ Period: _____

Directions: Go to the Web site your teacher writes on the board. You should see a live video of an animal. In this activity you will observe the animal's behavior over a three-day period. Record these observations twice a day for ten minutes at a time. Try to keep track of an individual animal if you can. For the Activity category, jot down what the animal is doing using details. For example, you may write, "the panda is sleeping in the back corner," or "the polar bear is pacing in front of the enclosure," or "the monkeys are playing by screeching and running toward each other."

Day One	Time	# of Animals	Color of animal	Activity
Observation 1				
Observation 2				

Day Two	Time	# of Animals	Color of animal	Activity
Observation 1				
Observation 2				

Life Science: Student Activity Page

Animal Cam (cont.)

Name: _____ Date: _____ Period: _____

Day Three	Time	# of Animals	Color of animal	Activity
Observation 1				
Observation 2				

Directions: Now use the information you gathered to answer the following questions:

1. Describe one behavior or activity an animal did that was unusual or that surprised you:

2. What did you learn about this animal?

3. Why do scientists make observations of animals? What do they hope to learn?

Human Body: Teacher Notes

Healthy Food Habits*

NSTA Standards (5–8):
A, F

Objectives:
Students will:
- brainstorm and discuss healthy and unhealthy foods
- complete the USDA food guide pyramid to learn about healthy eating habits

Suggested Web Site:
http://www.teachercreated.com/books/3842
Click page 66, site 1

Alternate Web Site(s):
Consumer Information Center: The Food Guide Pyramid
http://www.pueblo.gsa.gov/cic_text/food/food-pyramid/main.htm

Pre-Internet Activities:
- Direct students to brainstorm individually or in small groups the healthy and unhealthy foods they know, recording them on the sheet provided. Then ask students to briefly share their ideas, explaining for each food how they know it is healthy or unhealthy.
- Introduce the USDA Food Guide Pyramid by explaining that scientists have tried to determine the best diet for Americans, one that will supply all of the necessary nutrients that the human body needs. Based on scientific findings, the Food Guide Pyramid was created. The pyramid outlines the major food groups and recommends a certain number of servings per day from each food group.
- After the computer activity, instruct students to go back to their Healthy/Unhealthy Foods sheet and label each food listed by its food group according to the pyramid. For example, next to milk, students should write *dairy*.

Extensions:
- Ask students why they think the food guide is shaped as a pyramid. Allow students to examine their completed Food Guide Pyramids and discuss the question in groups to come up with an answer.
- Discuss the merits of the Food Guide Pyramid. Do students think it's the best possible diet recommendation? Direct students to try to find alternative food guides on the Internet, such as those that recommend a higher intake of nuts and "good" oils, such as olive oil. Alternatively, have students research the Japanese diet or the Mediterranean diet and compare it with the typical American's diet. Have scientists found positive aspects to the diets of other cultures? What are some simple changes Americans could make to their diets to be healthier?

Human Body: Student Activity Page

Healthy Food Habits *(cont.)*

Name: _____ Date: _____ Period: _____

Directions: Record all of the healthy and unhealthy foods you can think of in the chart below.

Healthy Foods	Unhealthy Foods

Human Body: Student Activity Page

Healthy Food Habits (cont.)

Name: _____ Date: _____ Period: _____

Directions: Go to **http://www.teachercreated.com/books/3842** and click page 66, site 1. Use the Web site to complete the blank Food Guide Pyramid below. Note: Be sure to label the food groups shown on the pyramid, including how many servings are recommended each day.

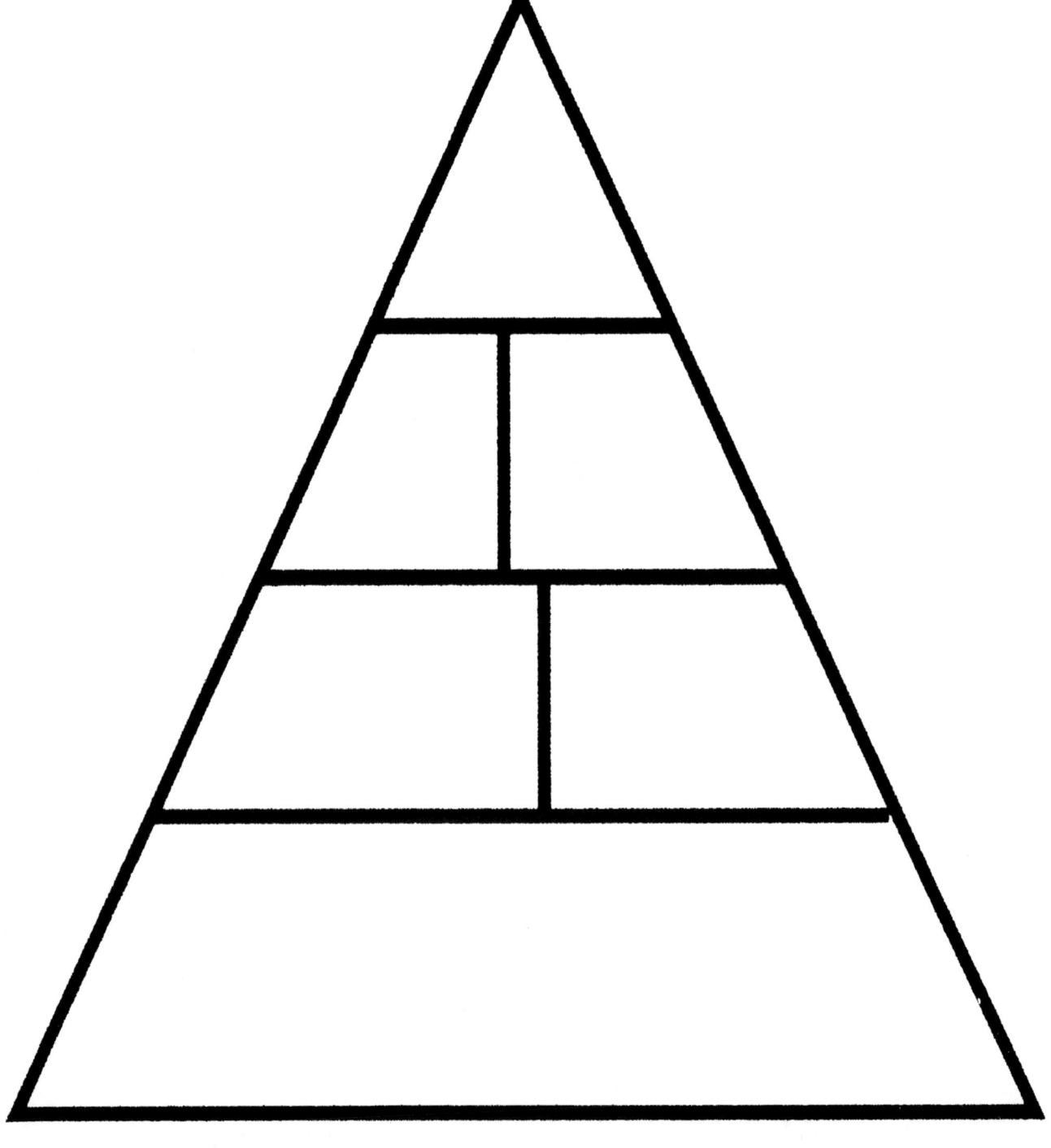

Human Body: Teacher Notes

Cells: Body Building Blocks*

NSTA Standards (5–8):
A, C

Objectives:
Students will:
- brainstorm the topic of cells
- understand that all organisms are composed of cells, which are the fundamental unit of life
- understand that many organisms are comprised of only a single cell, but humans and other organisms are comprised of many cells
- understand that cells sustain life by carrying out many functions
- draw a single-celled organism

Suggested Web Site:
http://www.teachercreated.com/books/3842

Click page 69, site 1

Alternate Web Site(s):
Enchanted Learning – Animal Cell Anatomy

http://www.enchantedlearning.com/subjects/animals/cell/

Pre-Internet Activities:
- In small groups or as a class, complete the first two columns on the KWL chart on the next page to brainstorm the topic of cells. When finished, read the following to the students, and instruct them to write these facts in the What have we Learned about cells column of their KWL charts (instruct them to leave some blank space below these facts for the computer activity):
    - all organisms are composed of cells, which are the fundamental unit of life.
    - most organisms are comprised of only a single cell. Humans and other organisms are comprised
      of many cells.
    - cells sustain life by carrying out many functions.

Extensions:
- Direct students to use the same Web site to learn what functions cells carry out. Instruct them to use this knowledge to write a few paragraphs about the functions of cells.
- Describe for students how specialized cells join together to form tissue, such as muscle. Tissue is then grouped together to form body parts, such as organs. Have students draw a simple diagram or flow chart illustrating this relationship.

Human Body: Student Activity Page

Cells: Body Building Blocks (cont.)

Directions: Brainstorm what you know and want to know about cells in the appropriate columns. Then listen to your teacher as she tells you more about cells. Record these facts in the What have we Learned about cells column. Then go to
http://www.teachercreated.com/books/3842
and click page 69, site 1. Use the Web site to read about cells. As you read and find interesting facts, record them in the remaining space in the What have we Learned about cells column.

What do we **Know** about cells?	What do we **Want** to know about cells?	What have we **Learned** about cells?

Use the same Web site to find and draw a picture of a single-celled organism in the box below:

Human Body: Student Activity Page

Body Systems*

NSTA Standards (5–8):
A, C, F

Objectives:
Students will:
- research two major body systems (Digestive, Respiratory, Reproductive, Circulatory, Excretory, Immune, Nervous, Muscular, or Skeletal) to determine the major parts and functions
- draw the body systems researched

Suggested Web Site:
http://www.teachercreated.com/books/3842

Click page 71, site 1

Alternate Web Site(s):
How the Nervous System Interacts with Other Body Systems
http://faculty.washington.edu/chudler/organ.html

Gander Academy's Human Body Systems
http://www.stemnet.nf.ca/CITE/body.htm

BBC – Science Human Body
http://www.bbc.co.uk/science/humanbody/

Pre-Internet Activities:
- Choose two body systems for the entire class to study or allow each student to choose two systems to study from the following list: Digestive, Respiratory, Reproductive, Circulatory, Excretory, Immune, Nervous, Muscular, or Skeletal.
- You may choose to research one system together as a class to show students how to determine the system's major parts and function(s).
- Make two copies of the Body Systems student activity page for each student.

Extensions:
- Have students who studied different systems work together to write an explanation of how two systems interact in the body. For best results, pair students whose systems interact in more obvious ways. For instance, pair students who studied the circulatory system with students who studied the respiratory system, and pair students who studied the muscular system with students who studied the skeletal system.
- Have students study how the body system they researched works. For example, if they studied the circulatory system, direct them to trace the path of blood as it flows through the system, and describe what each major part of the system does as it acts on the blood.

Human Body: Student Activity Page

Body Systems (cont.)

Name: _____ Date: _____ Period: _____

Directions: Go to **http://www.teachercreated.com/books/3842** and click page 71, site 1. Use the subject directory to research and draw two major body systems. In your drawing be sure to include and label all the major parts. Hint: human body systems is a subtopic of Anatomy, which is a subtopic of Biology.

Body system:	
Major parts:	Main function(s):

Drawing (be sure to label the major parts of the system):

TCM# 3842 Web Resources for Science Activities ©Teacher Created Materials, Inc.

Human Body: Teacher Notes

Exercise Choices**

NSTA Standards (5–8):
A, F

Objectives:
Students will:
- research and evaluate several forms of exercise using a comparison matrix
- understand the value of exercise to one's overall health
- research and understand the value of aerobic activity
- evaluate current personal exercise activities and create an exercise plan

Suggested Web Site:
http://www.teachercreated.com/books/3842
Click page 73, site 1

Alternate Web Site(s):
Kids Health – It's Time to Exercise!
http://kidshealth.org/kid/stay_healthy/fit/what_time.html

Pre-Internet Activities:
- Ask your students to discuss why exercise is important for their health and well-being. Invite students to share the types of exercise they do on a regular basis. Introduce and discuss a few unusual activities with which students might not be familiar, such as Pilates, spinning, tai-chi, etc.
- Point out all of the benefits of regular exercise such as maintaining a healthy weight, maintaining good overall physical health such as strong bones and muscles, maintaining healthy heart/lung systems, having energy for everyday activities, feeling good about yourself, and having a way to relieve stress and worries.

Extensions:
- Choose one of the following research projects (or allow students to choose): popular exercise activities in other countries, extreme-endurance sports, or unusual sports such as kite surfing. Allow students to research a broad area or choose just one sport. Then direct them to use the Internet to do their research. Specific aspects of sports students can research include the countries or parts of the U.S. where the sport is practiced, the equipment needed, the competitions, the physical benefits, and how the athletes train for the sport. When the students are finished gathering information, ask them to create a presentation (using a presentation software program if available) to share with the class.
- Discuss how to prevent injuries in a variety of sports. Your discussion can range from relatively safe activities such as running to more dangerous activities such as skateboarding.

Human Body: Student Activity Page

Exercise Choices (cont.)

Name: _____ Date: _____ Period: _____

Directions: Go to **http://www.teachercreated.com/books/3842** and click page 73, site 1. Use the Web site to find other sites about exercise. Using information from these Web sites, complete this page. Hint: type the keywords "aerobic exercise" or the specific exercise name such as "jogging" or "swimming" in the search engine search box.

1. What is aerobic exercise and why is it important? _____
   _____
   _____
   _____

2. Complete the matrix shown below by first choosing a type of exercise that is new to you (write its name below the words, "A new type of exercise") and then choosing another type of exercise (write its name below the words, "Another type of exercise"). Now research all 5 types of exercise listed in the matrix. Check the boxes that are true for each form of exercise.

Exercise Comparison Matrix

	It works major muscle groups	It's an aerobic exercise	I can do this exercise all year	It's enjoyable	It fits into my schedule	I have the equipment or facilities to do this exercise
Jogging						
Swimming						
Bike Riding						
A new type of exercise:						
Another type of exercise:						

Human Body: Student Activity Page

Exercise Choices (cont.)

Name: _____ Date: _____ Period: _____

Directions: On the first chart, record the exercise activities you do in a typical week. Remember to include any activity that gets your body moving, from walking to school to playing games in your P. E. class. Then answer the questions below the first chart. Finally, complete the last chart (further instructions are below).

Exercise I Do in a Typical Week

Sunday	Monday	Tuesday	Wednesday	Thursday	Friday	Saturday

Look at the chart you just completed to answer the following questions:

1. Write down all of the aerobic exercise you do here: _____
   _____
   _____

2. Do you do an aerobic exercise at least two to three times a week for 20-30 minutes at a time? (Yes or No.)

Directions: Use the information you learned from the comparison matrix on the previous page and from the chart above to create an exercise plan. If you are not already doing enough aerobic exercise, be sure to plan to do an aerobic exercise at least two to three times a week for 20-30 minutes. Also, try to fit a new form of exercise into your schedule just for fun.

Sunday	Monday	Tuesday	Wednesday	Thursday	Friday	Saturday

Human Body: Teacher Notes

Heredity: All About You***

NSTA Standards (5–8):
A, C

Objectives:
Students will:
- compare physical characteristics to a partner's, noting differences and similarities
- brainstorm how it is possible that children look similar to their parents.
- determine the meanings of several heredity vocabulary words and concepts
- label a diagram showing chromosomes, DNA, and genes

Suggested Web Site:
http://www.teachercreated.com/books/3842

Click page 76, site 1

Alternate Web Site(s):
Kids Only Genetics Glossary
http://www.genecrc.org/site/ko/kogloskids.htm#chromosome

KidsHealth: What is a Gene?
http://kidshealth.org/kid/talk/qa/what_is_gene.html

Genetic Science Learning Center: The Basics and Beyond
http://gslc.genetics.utah.edu/units/basics/

Pre-Internet Activities:
- Ask students to think about the people in their families and how they appear similar to one another. Does the sister have the same hair color as the mother? Does the brother have the same nose as the grandfather?
- You may choose to have the class work together complete the diagram showing chromosomes, DNA, and genes. After filling in the blanks, discuss each piece of the diagram and the role each piece plays in heredity.
- Note that due to the difficulty of the subject matter, you may choose to direct students to one of the alternate Web sites given rather than allow them to search for relevant Web sites on their own. Some of the Web sites may be easier to understand than others.

Definitions (answers to the word match):
DNA: The long spiral molecules that make up chromosomes. DNA contains ALL of the instructions for how to "build" a person, including what color eyes and hair a person will have, and how tall they will be.

Gene: A piece of DNA that defines a specific trait such as eye color or height. Thousands of genes make up DNA, which in turn make up chromosomes.

Chromosomes: The larger structures inside cells that contain DNA and genes.

Trait: A characteristic, such as eye color, skin color, or height. Traits are determined by genes.

Heredity: The passing of genes from one generation to the next.

Human Body: Student Activity Page

Heredity: All About You *(cont.)*

Name: _____ Date: _____ Period: _____

Directions: Compare your physical characteristics to a partner's physical characteristics using the chart below. Notice and record details (such as color, size, shape, etc.) about features like your eyes, hair, feet, height, skin, eyebrows, hands, nose, mouth, ears, etc.

Your Characteristics	Your Partner's Characteristics

Now circle the similarities and underline the differences between you and your partner.

Brainstorm: Have you ever wondered how people can have similar and different traits? For example, why do you look different from a friend but similar to your mother? Of course we know that we resemble our parents and even sometimes our grandparents, but what makes this possible? Brainstorm this last question with your partner and record your thoughts on the back of this paper.

Human Body: Student Activity Page

Heredity: All About You (cont.)

Name: _____ Date: _____ Period: _____

We look similar to our parents and grandparents because of heredity. Heredity is the process of passing characteristics (such as eye color, skin color, hair color, height, etc.) from one generation to the next. This happens during reproduction. When people, animals, and even plants reproduce, they create new individuals, and these individuals receive characteristics from both the male and the female parent. The female parent passes on her characteristics in her egg, and the male parent passes on his characteristics in his sperm. When egg and sperm unite, the instructions for the new individual are set. These instructions are contained in the genes of the new individual.

Now use the Internet to learn more about heredity. Follow the directions below.

Heredity Word Match

Directions: Go to **http://www.teachercreated.com/books/3842**
and click page 76, site 1.

Use the search engine or one of the Web sites listed on the board to complete the word match below. Hint: try typing the keywords, "heredity kids" or "genetics kids" in the search engine Search box to find relevant Web sites. Write the number of the word's definition next to the word.

Words	Definitions
_____DNA	1. Pieces of DNA that define specific traits such as eye color or height (Thousands of these pieces make up DNA, which in turn make up chromosomes.)
_____Heredity	2. A characteristic, such as eye color, skin color, or height
_____Trait	3. The long spiral molecules that make up chromosomes (These molecules contain ALL of the instructions for how to "build" a person, including what color eyes and hair a person will have, and how tall they will be.)
_____Chromosomes	4. The passing of genes from one generation to the next
_____Genes	5. The larger structures inside cells that contain DNA and genes

Human Body: Student Activity Page

Heredity: All About You (cont.)

Name: _____ Date: _____ Period: _____

Directions: Use the information you learned from the word match to label the diagram below. Feel free to refer back to the Web site(s) you visited previously to complete the diagram.

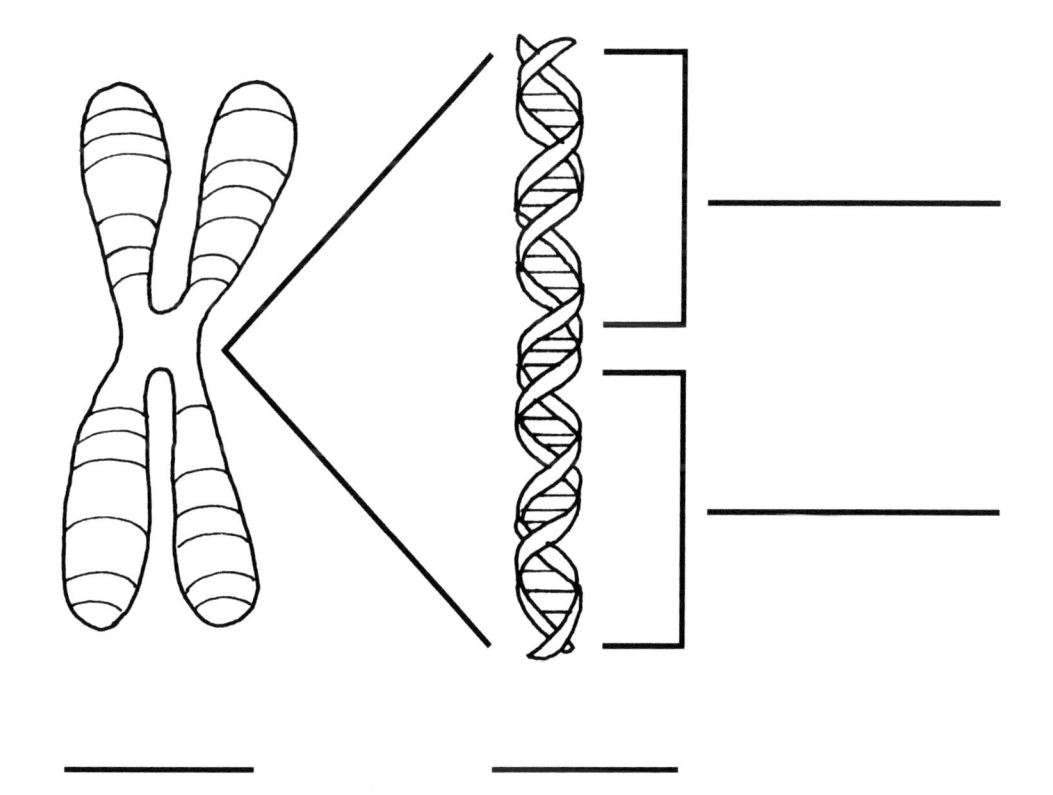

Bonus question: We have learned that genes determine how we look. However, have you ever heard people say things like, "Jimmy gets his baseball talent from his grandfather," or "Ally must get her musical ability from her father"? Do you think genes also determine our special abilities or talents? Are there other factors involved? Explain your answer.

Science in History and Society: Teacher Notes

Science News**

NSTA Standards (5–8):
A, F, G

Objectives:
Students will:
- find and read a science news article on the Internet
- extract and record the who, what, when, where, and why facts of the science article
- understand that science is constantly evolving and discoveries are made on an ongoing basis

Suggested Web Site:
http://www.teachercreated.com/books/3842

Click page 80, site 1

Alternate Web Site(s):
Science News for Kids

http://www.sciencenewsforkids.org

Pre-Internet Activities:
- Ask students if they have heard or read any science news recently. Perhaps they have heard of recent breakthroughs in disease research, new archaeological finds, recent discoveries in space, or even current animal news from a zoo. Discuss how we often hear about incredible science breakthroughs and discoveries. The fields of science are constantly changing as scientists make discoveries and build on each other's work.
- Explain to students that every good news article contains the who, what, when, where, and why facts of a current event. It is their job today to pull these facts out of a science news article they find on the Internet. Then they will form educated opinions about the science news event.

Extensions:
- Ask students to name the different fields of science. List them on the board. Then ask students to envision themselves as science reporters. For which field of science would they most like to report the news? Group students by common interest and direct them to imagine themselves as science reporters in their chosen field. Where would they most often go to gather the news? Would it be a research lab, an archaeological dig in another country, a university, or a zoo? Now have students create imaginary headlines for articles they might write as science reporters. If you have time, you might allow students to work together to write a science news article based on imaginary or actual events.

Science in History and Society: Student Activity Page

Science News *(cont.)*

Name: _____ Date: _____ Period: _____

Discoveries in science make front-page news almost every day. Newspaper and magazine readers are eager to read about the latest breakthroughs or discoveries in science. Science news can include breakthrough medicines that help fight disease, discoveries in space, or new computers that can perform amazing tasks. Science is constantly changing as we add new knowledge to our understanding of human beings, the earth, and the universe.

Directions: Now it's your turn to find an interesting science news article. Go to **http://www.teachercreated.com/books/3842** and click page 80, site 1. Use the Web site to find a science news article that interests you. Read it and record the who, what, when, where, and why facts of the article in the spaces below.

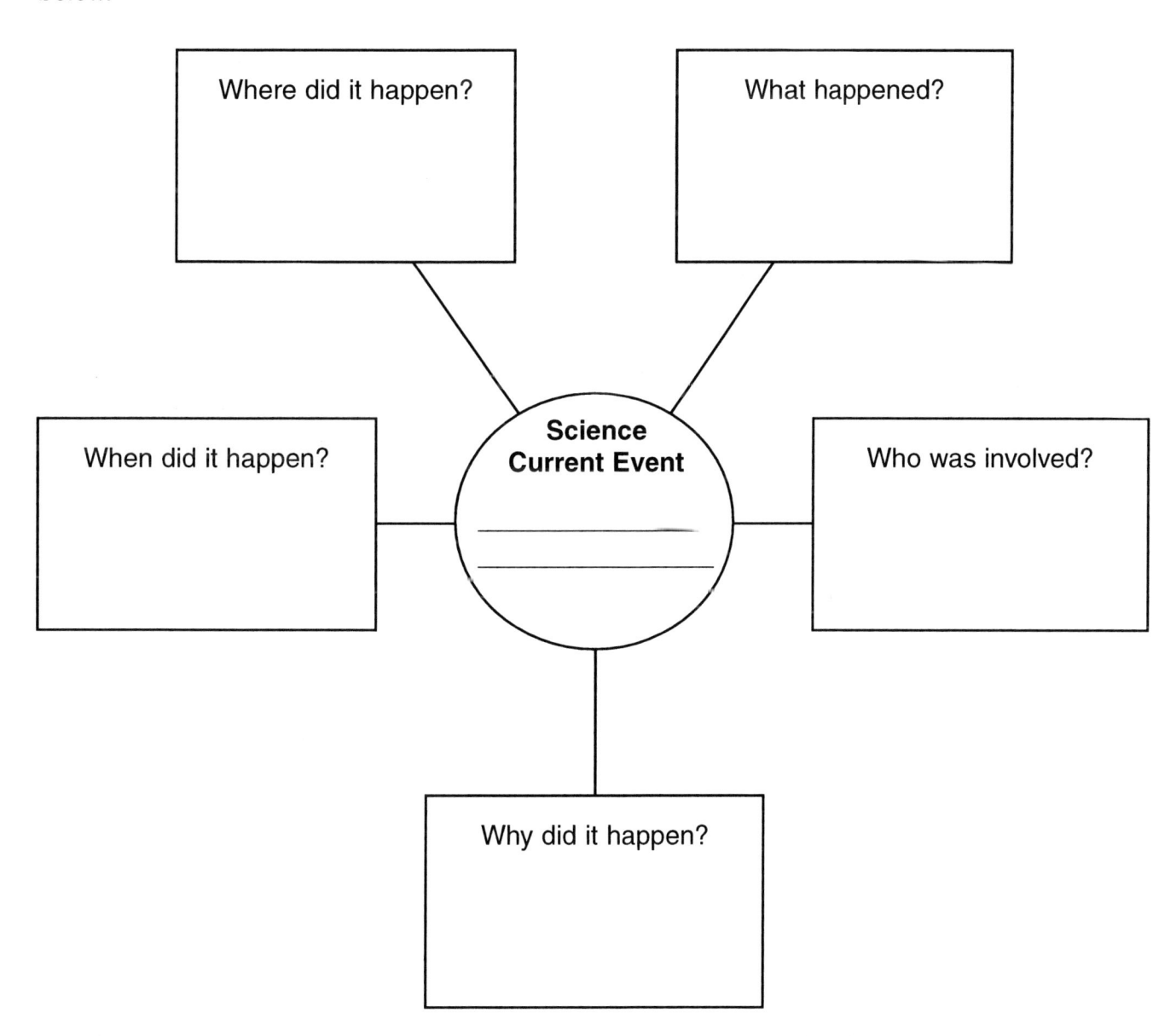

- Where did it happen?
- What happened?
- When did it happen?
- Who was involved?
- Why did it happen?

Science Current Event _____

Science in History and Society: Student Activity Page

Science News *(cont.)*

Name: _____ Date: _____ Period: _____

Directions: Using the science news article you found as well as the concept map you created, answer the questions below. Remember, some of these questions ask you to form an educated opinion. You won't necessarily find the answers to these questions in the article itself.

1. If your article was about a discovery, what kind of evidence did the scientist(s) provide to back up his or her discovery? Did it convince you that the discovery was real?

   _____
   _____
   _____
   _____

2. How do you think this science news will affect the work of other scientists in the same field?

   _____
   _____
   _____
   _____

3. How will this science news affect your life or someone else's life?

   _____
   _____
   _____
   _____
   _____
   _____
   _____

Science in History and Society: Teacher Notes

Fields of Science**

NSTA Standards (5–8):
A, E, F, G

Objectives:
Students will:
- research the major fields of science on the Internet
- describe or define several fields of science
- develop and investigate a question about each field of science

Suggested Web Site:
http://www.teachercreated.com/books/3842
Click page 83, site 1

Alternate Web Site(s):
Yahooligans Science and Nature
http://www.yahooligans.com/Science_and_Nature

Pre-Internet Activities:
- Ask students to name the major fields of science. Briefly discuss what scientists do in each field.

Extensions:
- Using the Internet, direct students to find a prominent scientist in the field that interests them most. Instruct students to find and record the scientist's educational background, discoveries and accomplishments, current place of employment, and if possible, what the scientist is currently researching. If desired, allow students to contact the scientist by e-mail and ask questions about his or her field or current research.
- Ask students to think about how different fields of science might overlap each other. Can they envision a scientific endeavor that might require scientists from different fields to work together? An example might be the quest to discover life in space, such as on Mars. Engineers, biologists, and astronomers work together to design and build the machines that will journey to Mars. Engineers focus on machine design and construction, biologists focus on the tests the machines must perform, and astronomers help to make sure the machines can survive the journey and actually get to Mars!
- Direct students to use the Internet to explain the focus of each of the following fields of science: Ichthyology, Meteorology, Volcanology, Aeronautics, and Seismology. Can the students find other less well-known fields of science?

Science in History and Society: Student Activity Page

Fields of Science (cont.)

Name: _____ Date: _____ Period: _____

Directions: Go to **http://www.teachercreated.com/books/3842** and click page 83, site 1. Use the Web site to look up each field of science listed below. Do this by typing each science field name in the Search box and pressing Enter. Choose an appropriate Web site from the list that appears. Write a definition for each field and a question that you have about that field. One field has been done for you.

- **Paleontology** — The study of...
- **Computer Science** — The study of machines and electronic systems that process information.
- **Astronomy** — The study of...
- **Physics** — The study of...
- **Biology** — The study of...
- **Chemistry** — The study of...
- **Geology** — The study of...

Fields of Science

Now think of a question about one of the fields above and try to answer it on the back of this page (either by using the Internet to look it up or by writing an educated opinion).

Science in History and Society: Teacher Notes

Famous Aviators & Technology***

NSTA Standards (5–8):
A, E, F, G

Objectives:
Students will:
- research famous aviators using the Internet
- record facts and data about aviators in a table
- examine and form opinions about one aviator's achievements in terms of historical significance, technological advances, and risk taking

Suggested Web Site:
http://www.teachercreated.com/books/3842
Click page 85, site 1

Alternate Web Site(s):
Yahooligans
http://www.yahooligans.com
Google
http://www.google.com

Pre-Internet Activities:
- Discuss the terms *aviation* and *aviators*. You may wish to use a KWL chart before beginning this activity and the following aviation activities. In the "K" or "What do we Know?" column, list all of the facts students know about aviation or flight. Then in the "W" or "What do we Want to know?" column, list all of the questions students have about aviation or flight. At the end of this mini-unit on aviation, brainstorm and list in the "L" column (the "What we Learned" column), what students learned.

- To prepare students for the second part of this activity, describe how airplane technology has come a long, long way. The first airplanes failed more often than they successfully took to the air. However, because aviators have tested and flown new aircraft, and because of advances in engineering, airplanes have become safer, faster, and more reliable.

- For the second part of this activity, it may be more appropriate for your students to answer the questions together as a class. However, if your students are ready, prepare them by discussing how to form an educated opinion. For practice, you might have them read about a military aircraft that was expensive to produce. After reading, ask them to give an educated opinion as to whether the aircraft was worth the cost.

Extensions:
- Have students discuss which aviator(s) they think contributed the most to our current knowledge of airplanes and flight. If students have different opinions, ask them to cite specific reasons and to do more research to back up their opinions. Arrange a classroom debate, with students of a similar opinion working together to make their case.

©Teacher Created Materials, Inc.

Science in History and Society: Student Activity Page

Famous Aviators & Technology (cont.)

Name: _____ Date: _____ Period: _____

Directions: Go to **http://www.teachercreated.com/books/3842** and click page 85, site 1. Use the Web site to research the following aviators. To research each aviator, type the aviator's name in the search engine Search box and press Enter. From the list that appears, click and view different Web sites until you find all the necessary information on that aviator. Repeat this process for the other aviators until you have completed the table.

Pilot Name	Date of Birth	Significant Flight(s) & Dates of each	Longest Distance Flown	Type or Name of Aircraft(s)
Wilbur or Orville Wright (circle one)				
Amelia Earhart				
Charles Lindbergh				
Chuck Yeager				

Science in History and Society: Student Activity Page

Famous Aviators & Technology (cont.)

Name: _____ Date: _____ Period: _____

Directions: Choose one of the aviators from the table on the previous page to investigate and think about further. Use the same search engine and Web sites to help you answer the following questions. Note that these questions require you to take the information you find and form an educated opinion.

1. Why was this aviator's most famous flight a breakthrough in aviation?

   _____
   _____
   _____
   _____

2. How did advances in airplane technology help this aviator to achieve his or her accomplishment?

   _____
   _____
   _____
   _____

3. Do you think this aviator took a big risk flying in the planes he or she flew?

   _____
   _____
   _____
   _____

4. If you answered yes to question #3, do you think the risk was worth taking? Did it benefit the aviator? Did it benefit other people, like you?

   _____
   _____
   _____
   _____

Science in History and Society: Teacher Notes

Preventable Natural Hazards?***

NSTA Standards (5–8):
A, F

Objectives:
Students will:
- define the term *natural hazard*, identify types of natural hazards, and learn when and where they occur
- research a type of natural hazard (such as a flood, wildfire, or landslide)
- identify human actions that can induce or exacerbate a natural hazard and/or damage from a natural hazard
- identify natural causes of a natural hazard

Suggested Web Site:
http://www.teachercreated.com/books/3842
Click page 88, site 1

Alternate Web Site(s):
FEMA for Kids: The Disaster Area
http://www.fema.gov/kids/dizarea.htm

National Geographic Nature's Fury
http://www.nationalgeographic.com/eye/natures.html

Pre-Internet Activities:
- Direct students to brainstorm the topic of natural hazards using the concept map on the next page. This concept map will direct students' brainstorming a bit by asking them to think about what constitutes a natural hazard (versus a structural or other man-made disaster), when and where they occur, and types of natural hazards. Invite them to add other major categories and thoughts to the concept map. When students are finished, invite them to share their thoughts, and come up with a class definition of natural hazards, identify all the types of natural hazards, and discuss when and where each type occurs. (See the next paragraph for this information.)
- A definition of a natural hazard: a physical event in the natural environment that causes damage to human life and human property. Types of natural disasters: earthquakes, landslides, wildfires, volcanic eruptions, floods, storms, and asteroid impacts. Where they occur: earthquakes and volcanic eruptions are most likely to occur at tectonic plate boundaries when these boundaries collide or move apart. Landslides are most likely to occur when there is heavy rainfall, where there are cliffs or steep slopes, and where there is a lack of vegetation to hold the soil in place. Wildfires are most likely to happen where there is dry or dead vegetation and extended periods of drought. As for storms, hurricanes happen in coastal areas in or near the tropics in summer and fall. Tornadoes need a combination of humidity, rising air, and heat near the ground, conditions which are more common in the southern U.S. and Midwestern U.S. in the spring. And finally, a large asteroid impact could happen anywhere on the globe. Scientists believe the last devastating asteroid impact wiped out the dinosaurs. As for future impacts, scientists are working on ways to detect and prevent these occurrences.

Science in History and Society: Teacher Notes

Preventable Natural Hazards? *(cont.)*

- Discuss ways in which human action can induce natural hazards or exacerbate the damage from natural hazards. The more obvious natural hazards to discuss in this context are floods, landslides, and wildfires. Floods naturally occur in floodplains near major rivers, where humans like to farm, create factories and other businesses, and live. Unfortunately, humans make flooding worse when they turn forested land near rivers into farmland, because trees and other vegetation they remove can help stem the floodwaters. Also, by building houses, factories, and farms in floodplains, human life and property is more vulnerable to damage should a flood occur.
- Direct students to complete the cause and effect graphic organizer by using the Internet to research the natural and human causes of wildfires, landslides, or floods. It is very difficult, if impossible, to find human causes for volcanic eruptions, earthquakes, and asteroid impacts. However, if your students need a difficult challenge, invite them to find possible human causes of devastating storms.

Extensions:

- Direct students to research natural disaster prevention efforts. Have students identify which natural disasters can be prevented all together (i.e., wildfires) and which ones we can prepare for in order to prevent extensive damage and/or loss of life. Have students choose one natural disaster to research (in the context of prevention efforts) in detail. Instruct them to explore the agencies and governmental organizations which work toward disaster preparedness and prevention. Finally, discuss or have students evaluate how human error in estimating the scope of a potential natural disaster can result in too much attention (and too much cost) being devoted to prevention efforts. Conversely, have students evaluate how too little preparation can result in significant human loss of life and property. An interesting discussion question would be, "How do scientific findings related to natural disasters influence the amount of attention and money devoted to natural disaster preparedness and prevention?"
- View satellite images of natural hazards on the Internet. At the time of this book's printing, this Web site was a good resource: EO Natural Hazards
  http://earthobservatory.nasa.gov/NaturalHazards/
- Invite students to research humanitarian responses to specific natural disasters. One resource is the ReliefWeb site sponsored by the United Nations Office for the Coordination of Humanitarian Affairs (OCHA)
  http://www.reliefweb.int/w/rwb.nsf/vLND

Science in History and Society: Student Activity Page

Preventable Natural Hazards? *(cont.)*

Name: _____   Date: _____   Period: _____

Directions: Use the concept map below to brainstorm the topic of natural hazards.

- When and where they occur
- What they are
- **Natural Hazards**
- Other facts/thoughts
- Types

Science in History and Society: Student Activity Page

Preventable Natural Hazards? *(cont.)*

Name: _____ Date: _____ Period: _____

Directions: Go to **http://www.teachercreated.com/books/3842** and click page 88, site 1. Use the search engine to research the human and natural causes of a natural disaster in order to complete the cause and effect diagram below. For the human causes, you can also list human actions that do not necessarily cause the natural disaster, but exacerbate its damaging effects. Note that it is best to choose floods, landslides, or wildfires. Hint: to find human causes of a particular type of natural disaster, try typing the keywords "human causes" and then the type of natural disaster. For example, "human causes wildfires."

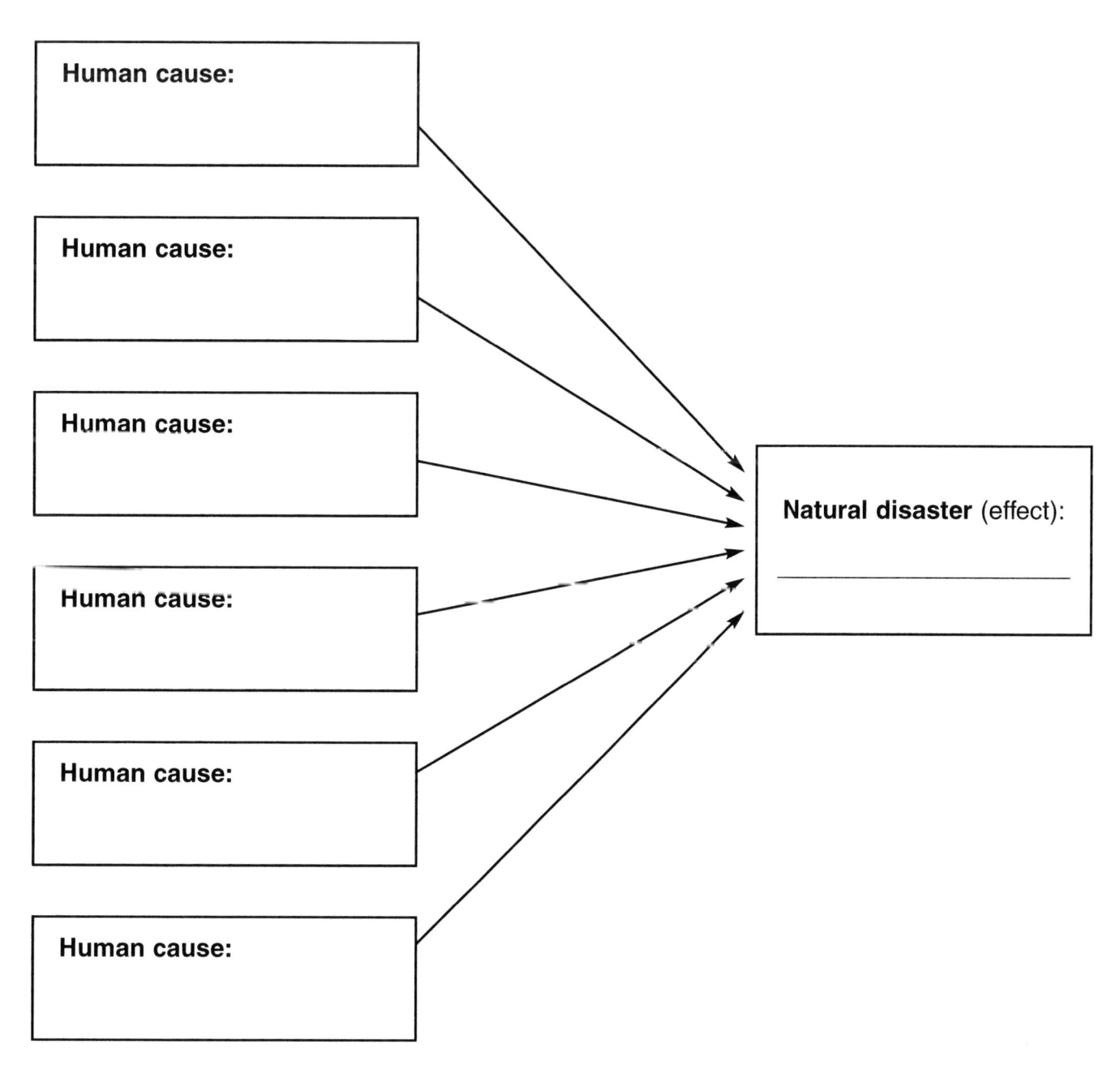

Human cause:

Human cause:

Human cause:

Human cause:

Human cause:

Human cause:

Natural disaster (effect): _____

Science in History and Society: Teacher Notes

Changes in Communication***

NSTA Standards (5–8):
A, F, G

Objectives:
Students will:
- learn when the telephone was invented and who invented it
- brainstorm and then research how people communicated before and after the invention of the telephone
- describe how the telephone has affected the ways that people act and interact
- describe how the telephone and other advances in communication technology have helped people conduct business
- determine how communication technology has impacted people's everyday lives
- evaluate how advances in communication technology have affected the student's own life

Suggested Web Site:
http://www.teachercreated.com/books/3842

Click page 92, site 1

Alternate Web Site(s):
Communication Before Telephones

http://www.angelo.edu/services/library/wtxcoll/verizon_web/pages/Timeline/pre_com/pre_com.htm

The History of Communication

http://inventors.about.com/library/inventors/bl_history_of_communication.htm

Pre-Internet Activities:
- Ask students to name several forms of communication technology such as telephones, televisions, radio, computers, fax machines, etc. Now ask students to imagine living in a world without these things. For most of human history, people have lived without such technology. How did people communicate?
- The telephone was patented by Alexander Graham Bell in 1876. Imagine how revolutionary and almost unbelievable such technology was to people of the time. It must have been very strange to hear someone's voice without being within earshot of that person! Imagine how incredible it was to be able to talk to someone in another building, in another city, and later, in another country (even across the ocean!).

Extensions:
- Invite students to dream up new forms of communication technology. What new invention would make it even easier for people to converse, send messages, or deliver information?

Science in History and Society: Student Activity Page

Changes in Communication *(cont.)*

Name: _____ Date: _____ Period: _____

Directions: Use the concept map below to brainstorm how people communicated before and after the invention of the telephone, including how people communicate today. After you are finished brainstorming, go to **http://www.teachercreated.com/books/3842** and click page 92, site 1. Use the search engine to find more ways people communicated before and after the invention of the telephone. Hint: to research ways people communicated before the telephone, try using the keywords "communication before telephones," "before the telephone," and "history of communication." To research ways people have communicated since the invention of the telephone, try the keywords, "20th century inventions" and "history of communication."

Communication

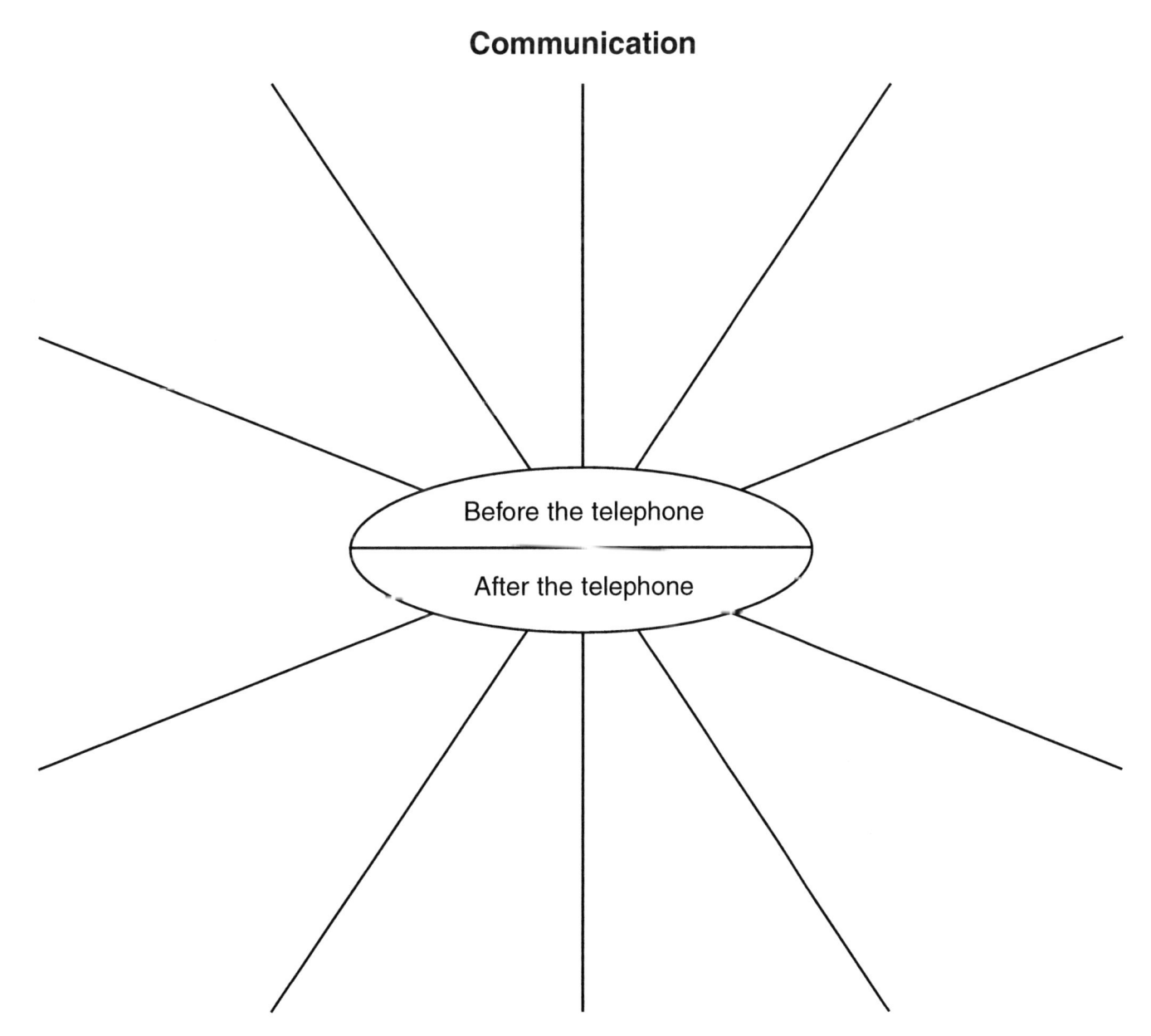

- Before the telephone
- After the telephone

Science in History and Society: Student Activity Page

Changes in Communication *(cont.)*

Name: _____ Date: _____ Period: _____

Directions: Now use the information you recorded on the concept map to answer the following questions. You may need to go back to the Web sites you visited earlier to help you answer these questions. Remember however, that for some of these questions, you will not be able to find the answers. Instead, you must offer an educated opinion.

1. How has the telephone affected the ways people act and interact?

   _____
   _____
   _____
   _____
   _____
   _____
   _____
   _____
   _____
   _____

2. How has the telephone helped people conduct business? How have other technological advances in communication helped people conduct business?

   _____
   _____
   _____
   _____
   _____
   _____
   _____
   _____
   _____
   _____

Science in History and Society: Student Activity Page

Changes in Communication *(cont.)*

Name: _____ Date: _____ Period: _____

3. Choose one other invention in communication technology (such as the radio, TV, computer, etc.) and imagine what life was like before the invention existed. Now describe how that invention has altered everyday life for most people.

_____
_____
_____
_____
_____
_____
_____
_____
_____

4. How is the quality of your life better because of advances in communication technology? Can you think of any ways in which the quality of your life is worse because of these advances in communication technology?

_____
_____
_____
_____
_____
_____
_____
_____
_____

Space Science: Teacher Notes

Packing for Space*

NSTA Standards (5–8):
A, D, E

Objectives:
Students will:
- find and explain the function of 10 items an astronaut uses on a space flight or on a space station
- develop an awareness of the unique requirements for survival in space

Suggested Web Site:
http://www.teachercreated.com/books/3842
Click page 96, site 1

Alternate Web Site(s):
StarChild
http://starchild.gsfc.nasa.gov/docs/StarChild/

Pre-Internet Activities:
- Briefly discuss NASA's space shuttle program and the International Space Station (ISS). Why do astronauts travel into space and what do they do there? Explain that astronauts aboard the space shuttle and space station conduct science experiments, make repairs to satellites and telescopes, launch new satellites and telescopes, and work on building more of the ISS. Introduce the activity by asking, *given that astronauts have work to do, but also have to live and survive in space, what do you suppose they need to take with them when they travel into space?*

Extensions:
- Ask students to brainstorm what they would need to take with them if they were to travel into space. Would they need anything different than what the astronauts take? Remind students that there is limited space on the space shuttle and in the ISS. Have each student write their personal packing list with an explanation next to each item as to why they need to take it along.

Packing List Possible Answers:

Launch and Entry Suit	Experiment equipment	Manned Maneuvering Unit
Space suit	Experiment specimens	Extravehicular Maneuvering Unit
Exercise shorts & t-shirts	Tools	Sleeping bag
Work shirts & pants	Computers	Astronaut food
Sweaters/pullovers	Playing cards	Water
Work & exercise shoes	Books	

Space Science: Student Activity Page

Packing for Space (cont.)

Name: _____ Date: _____ Period: _____

Directions: Imagine that you are an astronaut preparing for a space flight that will take you to live on a space station. What do you need to take with you to survive and work in space? Go to **http://www.teachercreated.com/books/3842** and click page 96, site 1. Use the Web site to find 10 items that an astronaut uses or needs in space. After each item describe how an astronaut uses it to survive or work in space.

1. Item: _____

   How it is used: _____

2. Item: _____

   How it is used: _____

3. Item: _____

   How it is used: _____

4. Item: _____

   How it is used: _____

5. Item: _____

   How it is used: _____

6. Item: _____

   How it is used: _____

7. Item: _____

   How it is used: _____

8. Item: _____

   How it is used: _____

9. Item: _____

   How it is used: _____

10. Item: _____

    How it is used: _____

Bonus question: Why don't you see pictures of astronauts inside the space shuttle wearing spacesuits? (Write your answer on the back of this page.)

Space Science: Teacher Notes

The International Space Station**

NSTA Standards (5–8):
A, D, E, F, G

Objectives:
Students will:
- describe the work of the astronauts living on the ISS
- identify countries involved in building the ISS and their contributions
- analyze the importance of the ISS

Suggested Web Site:
http://www.teachercreated.com/books/3842

Click page 98, site 1

Alternate Web Site(s):
Discovery.com: International Space Station
http://www.discovery.com/stories/science/iss/iss.html

NASA Human Space Flight International Space Station
http://spaceflight.nasa.gov/station/

Pre-Internet Activities:
- Discuss what students know about the International Space Station (ISS). Briefly describe how the ISS is a very large "house" orbiting the earth. It is where astronauts live and work. Several countries have worked together to build the ISS piece by piece, and they are still working together to complete it. Aboard the ISS astronauts conduct science experiments, taking advantage of the unique micro-gravity environment. Astronauts also have an important view of earth from the ISS. This view helps us understand weather changes, geological events, and environmental problems.

Extensions:
- You may wish to structure the above pre-Internet activity using a KWL chart. In the "K" or "What do we Know?" column, list all of the facts students know about the ISS. Then in the "W" or "What do we Want to know?" column, list all of the questions students have about the ISS. Finally, after students have read about the ISS and completed The International Space Station activity sheet, brainstorm and list in the "L" column (the "What we Learned" column) what students learned.
- Challenge students to understand the political issues of the ISS. Did all of the countries involved readily join the effort in building the ISS, or was there resistance from some countries? What countries are not involved and why aren't they? Do any countries have political motives for being a part of the ISS? Do students think there are any countries that wish they could build their own space station (and not be a part of the U.S.-led ISS)?

Space Science: Student Activity Page

The International Space Station *(cont.)*

Name: _____ Date: _____ Period: _____

Directions: Go to **http://www.teachercreated.com/books/3842** and click page 98, site 1. Use the Web site to read about the International Space Station (ISS). Write the answers to the questions below in the spaces provided.

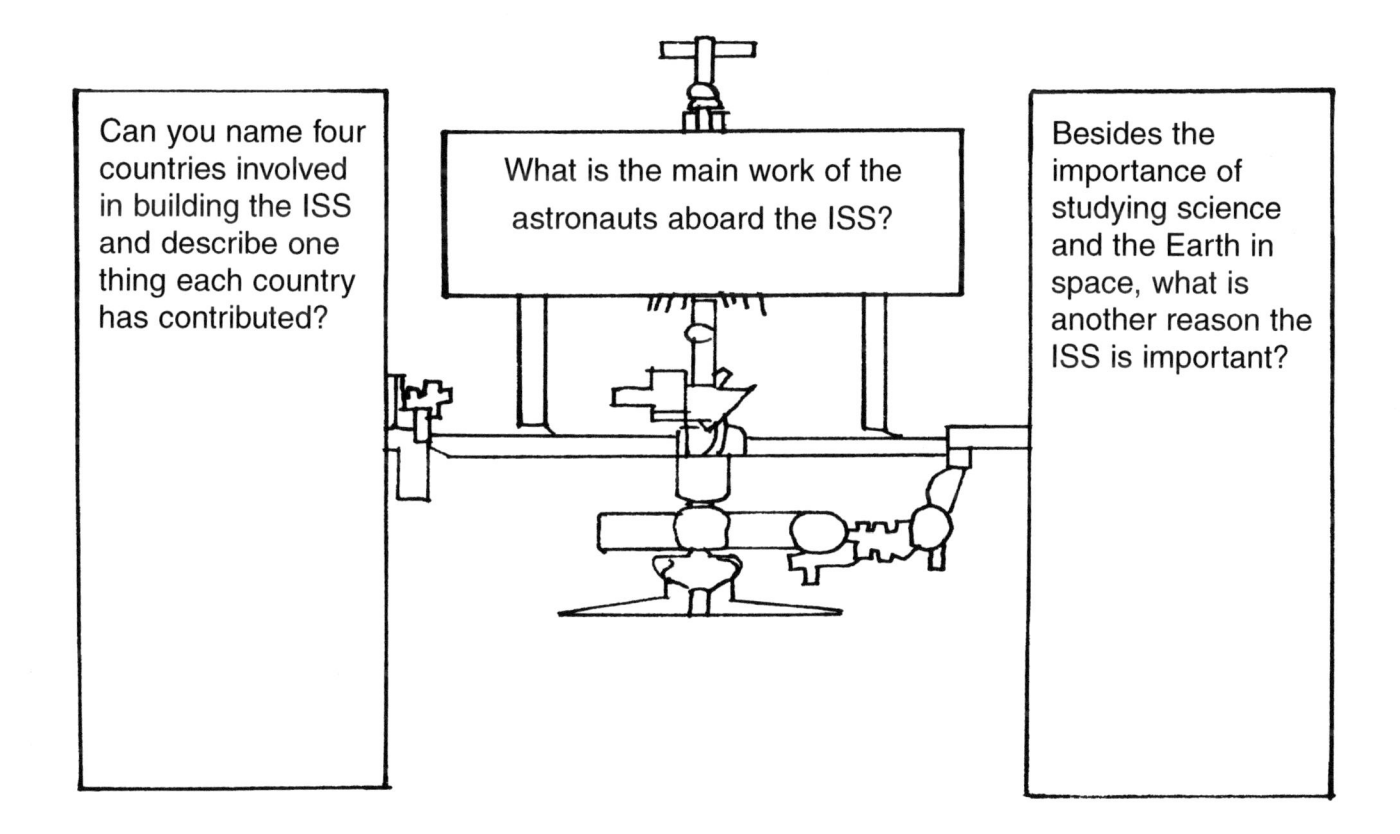

Can you name four countries involved in building the ISS and describe one thing each country has contributed?

What is the main work of the astronauts aboard the ISS?

Besides the importance of studying science and the Earth in space, what is another reason the ISS is important?

Space Science: Teacher Notes

Phases of the Moon***

NSTA Standards (5–8):

A, D

Objectives:

Students will:

- visualize what the moon looks like from earth using a diagram showing how the sun shines on the moon as it travels around the earth
- use a search engine to find what the moon looks like today
- name the current phase of the moon
- use a search engine to find the moon's rise and set times for today as viewed from the student's hometown

Suggested Web Site:

http://www.teachercreated.com/books/3842

Click page 100, site 1

Alternate Web Site(s):

What the Moon Looks Like Today: Apparent Disk of Moon

http://aa.usno.navy.mil/idltemp/current_moon.html

Complete Sun and Moon Data for One Day

http://aa.usno.navy.mil/data/docs/RS_OneDay.html

Pre-Internet Activities:

- Ask students if they know why the moon sometimes appears as a crescent, sometimes appears full, and sometimes cannot be found in the sky at all. Brainstorm the reasons why this might be so. Then ask students if the moon follows any kind of pattern. Allow students to share what they know about the phases of the moon.
- Tell students that the moon does follow a definite pattern every 29.5 days, or roughly every month. This is how long it takes for the moon to orbit the Earth, and as it orbits the Earth, it takes on different shapes from our point of view. The reason for this is the sun's illumination. The sun shines on the moon all the time, but as it orbits the Earth its appearance changes. Show students the diagram on the first Phases of the Moon activity sheet to illustrate this point.
- Note: Allow students to complete the first Phases of the Moon student activity sheet in pairs, as the visualization activity can be difficult.

Extensions:

- Direct students to make nightly observations of the moon and record what they see both in drawing form and in writing. Have them observe for a month in order to view the entire cycle of the moon's phases. Then discuss what it means for the moon to rise and to set and how this relates to the sunrise and the sunset.

Space Science: Student Activity Page

Phases of the Moon *(cont.)*

Name: _____ Date: _____ Period: _____

Directions: The diagram below shows how the sun illuminates the moon as it orbits the Earth. Match the numbers shown in the first diagram to the drawings of the moon's phases in the second diagram. In order to do this, imagine how the moon would look to you if you were standing on the Earth in the first diagram.

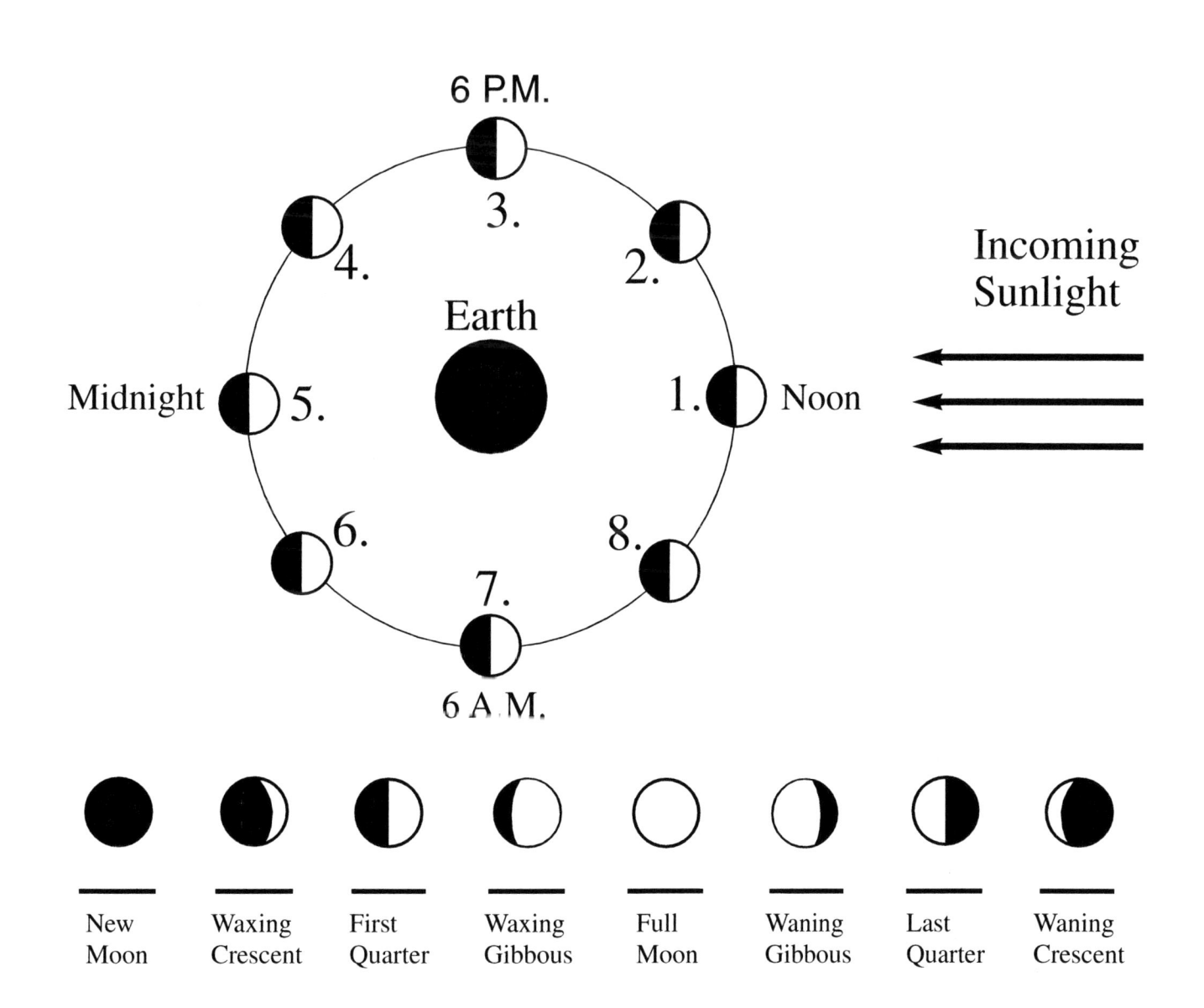

6 P.M.

3.

4.   2.

Earth

Midnight   5.   1.   Noon

Incoming Sunlight

6.   8.

7.

6 A.M.

New Moon	Waxing Crescent	First Quarter	Waxing Gibbous	Full Moon	Waning Gibbous	Last Quarter	Waning Crescent

©Teacher Created Materials, Inc.   TCM# 3842 Web Resources for Science Activities

Space Science: Student Activity Page

Phases of the Moon (cont.)

Name: _____ Date: _____ Period: _____

Directions: Go to **http://www.teachercreated.com/books/3842** and click page 100, site 1. Use the subject directory to find Web sites about the moon. Pick and choose among the Web sites to find the answers to the following questions. Hint: the Moon is a subtopic of the Solar System, which is a subtopic of Astronomy and Space.

1. Find a Web site that shows a picture of what the moon looks like today. Draw it here:

   Using the second diagram on the first *Phases of the Moon* student activity page, name the current phase of the moon (such as New Moon, Waxing Crescent, Last Quarter, etc.):
   _____

2. Find a Web site that shows the times that the moon will rise and set today where you live. On the Web site, type the information required such as your city name and the date, and then click the search button.

   What time will the moon rise today? _____

   What time will it set? _____

Bonus question: What does waxing and waning mean when describing the moon's phases?

Space Science: Teacher Notes

Earth is Unique**

NSTA Standards (5–8):
A, D, E

Objectives:
Students will:
- compare the earth's appearance and surface features to other planets using satellite pictures
- determine how the Earth's surface is unique
- understand that scientists use advanced satellite and telescope technology to learn about the planets of our solar system

Suggested Web Site:
http://www.teachercreated.com/books/3842
Click page 103, site 1

Alternate Web Site(s):
Google
http://www.google.com
(Search for "solar system pictures")

Pre-Internet Activities:
- Ask students if they know how scientists learn about the other planets in our solar system. Discuss ideas, and if no one suggests it, mention the pictures that come from satellites. Powerful telescopes mounted on satellites can take detailed pictures of planets, stars, asteroids, and even far-away galaxies. In fact, the Hubble Space Telescope has provided new glimpses into the farthest reaches of space in just the last few years. Just as satellites can provide detailed images of Earth (see the lesson *Our World from Space*), they can also provide the best images taken to date of the other planets and deep space.
- Provide colored pencils for students to draw pictures of the planets.

Extensions:
- Study the Hubble Space Telescope and some of the spectacular images it has provided of deep space. Type "hubble space telescope images" in a search engine Search box to find these images. Have students report on particular images to broaden everyone's understanding of the structures in space. Or, you might want students to study the history of the Hubble telescope, its technology, how it has expanded scientists' understanding of space, or the repairs astronauts have carried out to ensure its success.
- Study the scientific quest to determine if Mars has ever harbored life. Explore how satellite images have helped astronomers form theories that both support and discredit the argument that Mars once sustained life. Examine NASA's efforts to send rovers to Mars to study the planet's surface and determine if life once existed there.

Space Science: Student Activity Page

Earth is Unique (cont.)

Name: _____ Date: _____ Period: _____

Directions: Go to **http://www.teachercreated.com/books/3842** and click page 103, site 1. Use the subject directory to find pictures of the planets in our solar system. Then complete the exercises below. Hint: the Solar System is a subtopic of Astronomy and Space.

1. Pretend that you are an astronomer who wants to determine how the Earth is different from other planets. You are excited by new satellite pictures taken of each of the planets. You think these images will help you to "look" at each planet's surface and compare it to Earth.

   To start, you need to look closer at the planet Earth. Find a satellite picture of the planet Earth that shows some of its features such as land masses or landforms, and draw it here:

   Earth

On the back of this page, describe the Earth's surface as seen in the picture you drew (describe colors, land forms such as mountains, amount of water, continents, etc.)

Space Science: Student Activity Page

Earth is Unique *(cont.)*

Name: _____ Date: _____ Period: _____

2. Now use the same subject directory find a satellite picture of another planet that shows Some of its features. Draw it here:

   +--+
   | **Name of planet:** _____ |
   | |
   | |
   | |
   | |
   | |
   | |
   | |
   | |
   +--+

   Describe the planet's surface as seen in the picture you drew (describe colors, land forms such as craters, valleys, mountains, or other distinctive features like the red spot on Jupiter) :

   _____

   _____

   _____

3. Now after studying each planet, what do you think makes the Earth unique in terms of its features?

   _____

   _____

   _____

Space Science: Teacher Notes

Our Solar System*

NSTA Standards (5–8):

A, D

Objectives:

Students will:

- label a diagram of the solar system
- discover that the Earth is the third planet from the sun, the sun is the largest body in our solar system and is a star, and there are nine planets in our solar system

Suggested Web Site:

http://www.teachercreated.com/books/3842

Click page 106, site 1

Alternate Web Site(s):

Solar System Printout/Coloring Page: Enchanted Learning.com

http://www.enchantedlearning.com/subjects/astronomy/activities/coloring/Solarsystem.shtml

Yahooligans Subject Directory

http://www.yahooligans.com

Click **Science and Nature, Astronomy and Space**, then **Solar System**.

Pre-Internet Activities:

- Use the provided concept map to help students brainstorm what they know about our solar system. They should record any thoughts that come to mind. You can pair students for this activity or allow them to work individually.

Extensions:

- Find directions on the Internet for making a solar system model and then do this activity with your students. Try typing "solar system model" in the Search box of a search engine such as Google to find the directions.
- After students have labeled their diagrams, allow them to check each other's work. Then, use the diagrams to discuss the orbits of the planets (or go online to find a site that shows the orbits more clearly. Try typing, "planet orbits" in a search engine Search box to find diagrams of the planets' orbits). Point out how the orbits are regular and predictable, just like the earth's orbit. Note that the regularity of the earth's orbit explains why the earth has regular and predictable seasons. Ask students to predict whether the other planets experience seasons like the earth.
- Discuss how distance from the sun affects the planets. Ask students to predict which planets are the coldest and which are the hottest. Do students think any other planets could be habitable based on their temperatures?

Space Science: Student Activity Page

Our Solar System (cont.)

Name: _____ Date: _____ Period: _____

Directions: Use the concept map below to brainstorm what you know about the solar system.

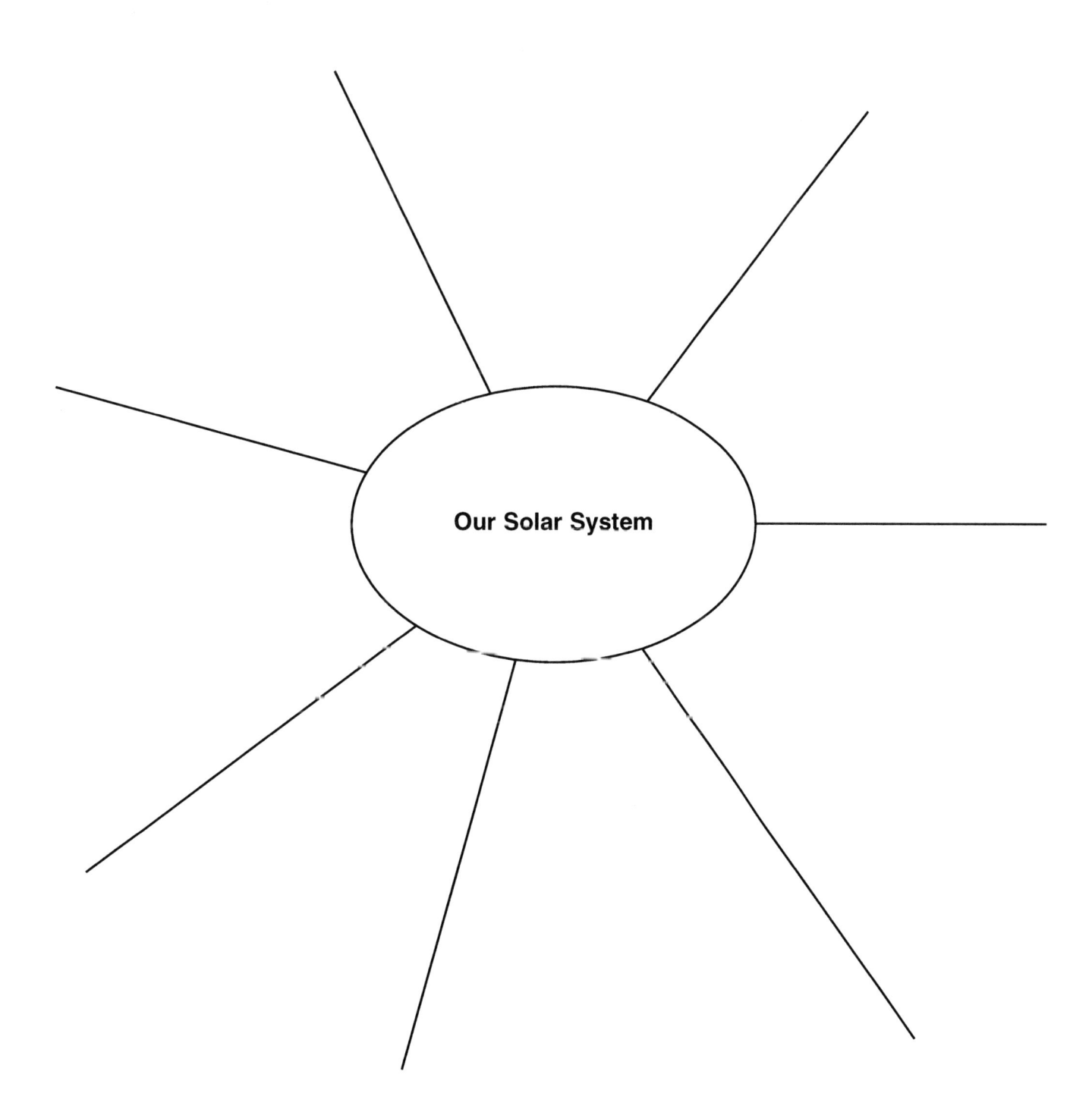

Our Solar System

Space Science: Student Activity Page

Our Solar System *(cont.)*

Name: _____ Date: _____ Period: _____

Directions: Go to **http://www.teachercreated.com/books/3842**
and click page 106, site 1. Use the search engine to label the solar system diagram below.
Hint: try typing the keywords, "solar system diagram" in the search engine Search box.

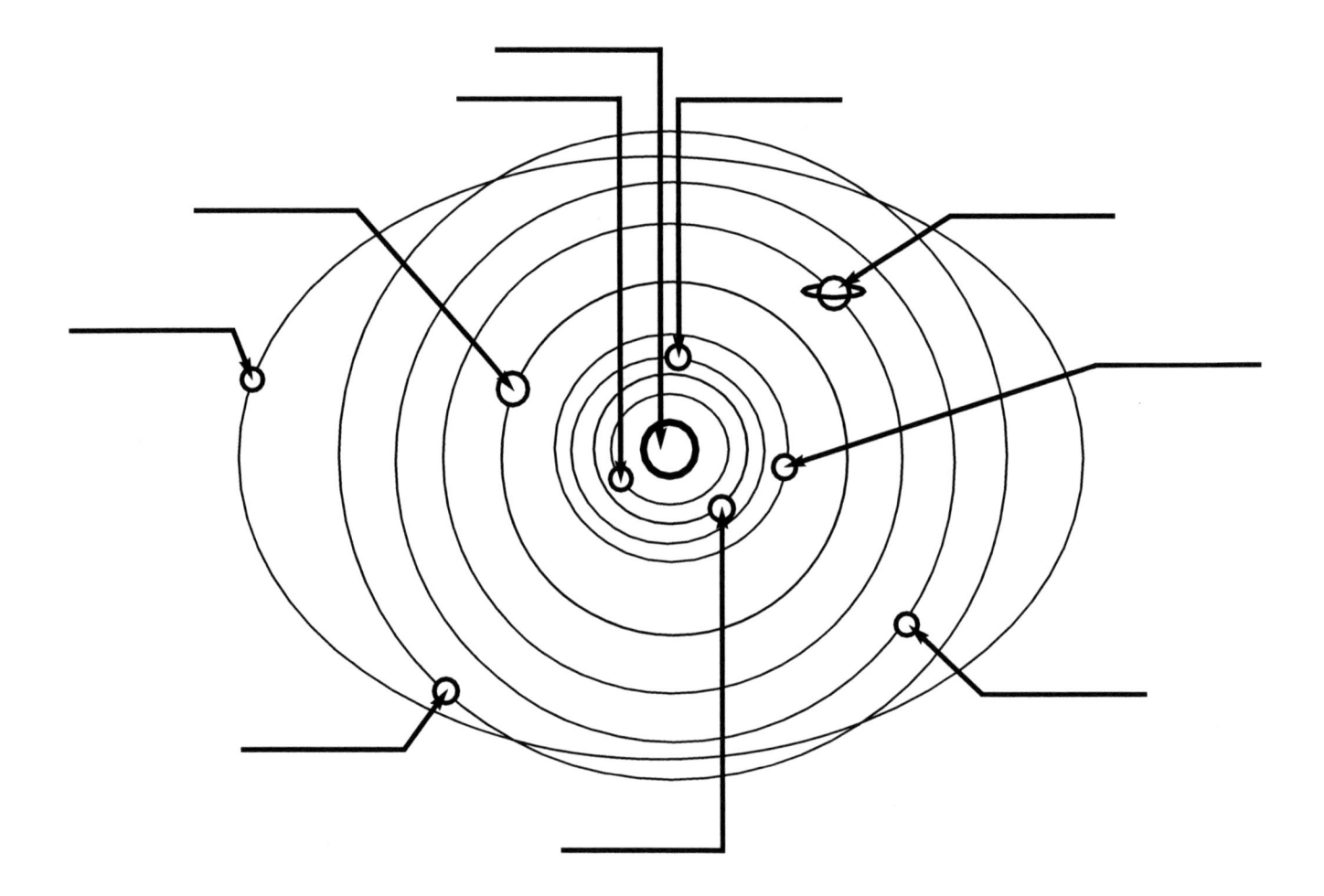

Now use the diagram to answer the following questions:

1. The Earth is how many planets away from the sun? _____

2. The largest body or object in our solar system is a star called the _____.

3. How many planets all together are in our solar system? _____

Bonus question: What is the name of the force that keeps planets in orbit around the sun?
Hint: this force also holds objects to the earth (like you and me!). _____

Space Science: Teacher Notes

Constellations & Myths*

NSTA Standards (5–8):
A, D, G

Objectives:
Students will:
- find a constellation online and draw it as it appears in the night sky
- locate a story or myth told about the constellation and paraphrase it

Suggested Web Site:
http://www.teachercreated.com/books/3842
Click page 109, site 1

Alternate Web Site(s):
Myths about the Sky, Constellations, and Stars: University Corporation for Atmospheric Research
http://www.windows.ucar.edu/tour/link=/mythology/stars.html

Pre-Internet Activities:
- Ask students if they have ever spent a lot of time looking at the night sky. Allow students to share what they've seen. Now ask if they have ever looked at the stars and imagined seeing the outlines of creatures and objects there. Long ago, when there was no TV or city lights, people used to see a lot more of the sky. The stars shone very brightly night after night. People got to know the patterns of the stars, the moon, and other objects very well. They saw the outlines of people, animals, and other objects in the stars. They made up stories to explain the things they saw in the night sky.

Extensions:
- For homework, challenge students to find the constellation they drew in the night sky. (Note however, that depending on the time of year, their constellation may or may not be viewable. You may decide to create a homework sheet that shows a constellation that is currently viewable. On the sheet, draw the constellation and tell where to find it in the sky.)
- This is a fun online constellation activity (this Web site was current at the time of the book's printing): Match the Constellations: ss_flashcards
  http://www.knowble.com/land/observatory/constellations.html
- Direct students to study one constellation in detail, researching different versions of the constellation's myths, the names of the stars that make up the constellation, how far away the stars are, etc. Or, have them compare and contrast how different cultures have viewed star patterns. Are there any star patterns that very different societies (i.e., Greek and African) have identified in the same way? If so, are the myths similar or very different? Another research idea is this: What other ways have various cultures tried to explain objects (i.e., comets, planets, etc.) in the night sky?

Space Science: Student Activity Page

Constellations & Myths *(cont.)*

Name: _____ Date: _____ Period: _____

Long ago, people looked up at the night sky and saw shapes, people, and animals in the stars. They made up stories about the things they saw. For example, the Greeks told a story about the constellation Orion. Orion was an excellent hunter who boasted that he could kill any animal on earth. Hearing this challenge, a small scorpion fought him, stung him, and killed him. Now Orion is in the heavens, and he stays well ahead of the scorpion as he hunts in the night sky.

Directions: Go to **http://www.teachercreated.com/books/3842**
and click page 109, site 1. Use the search engine to find a Web site that shows constellations. To do this, type "constellations" in the search engine Search box. Choose a Web site that shows you what constellations look like in the sky. Pick one constellation and draw it in the box below.

Constellation name: _____

Space Science: Student Activity Page

Constellations & Myths *(cont.)*

Name: _____ Date: _____ Period: _____

Directions: Use the same search engine (go to **http://www.teachercreated.com/books/3842** and click page 109, site 1) to research the constellation you drew on the previous page. To do this, type the name of the constellation in the search engine Search box and press Enter on your keyboard. Choose the Web site that you think will tell you the constellation's story or myth. Read the story and write it in your own words below.

Constellation name: _____

The constellation's story:

_____

_____

_____

_____

_____

_____

_____

_____

_____

_____

_____

_____

_____

_____

_____

_____

_____

_____

©*Teacher Created Materials, Inc.*   TCM# 3842 *Web Resources for Science Activities*

Space Science: Teacher Notes

Lights in the Northern Sky***

NSTA Standards (5–8):
A, B, D, G

Objectives:
Students will:
- compose a non-scientific explanation for the Northern Lights
- compose a scientific explanation of the Northern Lights
- determine where it is best to view the Northern Lights

Suggested Web Site:
http://www.teachercreated.com/books/3842
Click page 112, site 1

Alternate Web Site(s):
NORDLYS – Northern Lights
http://www.northern-lights.no/

Pre-Internet Activities:
- Ask students if they have ever seen the Aurora Borealis or Northern Lights. Let students share their experiences. If no one has seen them, explain that the Northern Lights are a colorful and magnificent display of light in the night sky. Tell students that they will see this phenomenon online today.

Extensions:
- Tie this lesson into a unit on magnetism or light.
- Ask students to research some of the ways ancient people explained the phenomenon of the Northern Lights. You might decide to assign different ancient cultures (such as the Vikings, the Danish, and Native Americans) to groups of students. When students are finished with their research, instruct them to create presentations and present to their fellow students.
- Instruct students to research and understand why it is difficult to predict the Northern Lights. Then challenge them to determine how people go about trying to predict them. For fun, tell students to pretend that they have come up with a way of predicting the Northern Lights. What would they do with this knowledge? Who do they think would be interested?
- Have students determine if other planets have auroras. If so, are they caused by the same factors on other planets? What are the key ingredients for a planet to have an aurora? What do scientists think they look like on other planets? How are they different than earth's Northern Lights and why?

Space Science: Student Activity Page

Lights in the Northern Sky *(cont.)*

Name: _____ Date: _____ Period: _____

Directions: Go to **http://www.teachercreated.com/books/3842** and click page 112, site 1. Use the search engine to find a Web site that contains a movie showing the Northern Lights. To do this, type "Aurora Borealis movie" or "Northern Lights movie" in the search engine Search box. Choose the best Web site and view the movie.

1. Imagine someone looking at the Northern Lights five hundred years ago, before science could explain the aurora. This person sees strange lights in the sky and wonders what they are. What sort of explanation do you think this person will make for the phenomenon he has seen?

   _____
   _____
   _____
   _____
   _____

Directions: Use the Google search engine (**http://www.google.com**) to find a Web site that explains the Northern Lights (and where it is best to view them) in scientific terms. Read the explanation and answer the questions below.

2. In scientific terms, what causes the Northern Lights?

   _____
   _____
   _____
   _____
   _____

3. Where is viewing of the Northern Lights the best?

   _____
   _____
   _____
   _____
   _____

Science and Technology: Teacher Notes

Inventing the Flying Machine***

NSTA Standards (5–8):
A, E, F, G

Objectives:
Students will:
- discover that nature inspires technological innovation
- describe how invention is a process of trial and error
- understand that there is no such thing as perfect technology. All inventions can be improved upon in safety, cost, efficiency, and/or appearance.
- evaluate an early aviator's success at inventing a flying machine
- determine how one inventor's successes and failures help other scientists in their scientific and technological quests
- compare early forms of aircraft to modern forms of aircraft

Suggested Web Site:
http://www.teachercreated.com/books/3842

Click page 114, site 1

Alternate Web Site(s):
Otto Lilienthal MUSEUM

http://home.t-online.de/home/LilienthalMuseum/ehome.htm

Pre-Internet Activities:
- Take a walk outside. Have the students sit quietly, and ask them to imagine themselves as inventors, trying to find a way for people to fly. What might they observe for ideas on how to fly? Birds, of course! Discuss what early scientists and inventors might have learned from birds. Perhaps they studied birds' wings: how they flap to create lift, and how they move to turn the bird in the air. Perhaps early inventors tried to create wings for people!
- Explain to students that in the early days of human flight, many people experimented with different aircraft in trail-and-error attempts to be the first to "fly." Many of these aircraft never made it off the ground. However, with each success, inventors improved upon aircraft design. Today we have amazing aircraft that can carry tons of cargo, fly at supersonic speeds, and carry astronauts into space and back. However, as incredible as our airplanes are, they can always be improved.

Extensions:
- Have students use the template on page 117 to make their own airplanes. Use cardstock or tag board for best results. As an added challenge, direct students to improve upon the design of this plane (allowing students to work in teams may produce better results). Specify that they must design a plane that will perform a specific function, such as travel far, carry weight (such as paperclips), fly in loops, etc. You might wish to set aside a special day when they can test their planes outside, and then award prizes in different categories.

Science and Technology: Student Activity Page

Inventing the Flying Machine (cont.)

Name: _____ Date: _____ Period: _____

Directions: Otto Lilienthal of Germany was an early inventor of flying machines. You can investigate his attempts to fly by using a search engine. Go to **http://www.teachercreated.com/books/3842** and click page 114, site 1. Type his name in the search box and press Enter. View Web sites from the list and read about Otto Lilienthal until you find the answers to these questions.

1. Describe how Otto Lilienthal's attempts at inventing a flying machine were a process of trial-and-error:

   _____
   _____
   _____
   _____
   _____

2. Do you think Otto Lilienthal was successful in what he was trying to achieve? Why or why not?

   _____
   _____
   _____
   _____
   _____

3. How did his successes and failures help other inventors across the world to design and test flying machines? (**Hint:** if you cannot find the answer to this question, give your opinion instead.

   _____
   _____
   _____
   _____

Science and Technology: Student Activity Page

Inventing the Flying Machine *(cont.)*

Name: _____ Date: _____ Period: _____

4. Draw a picture of one of Otto Lilienthal's later gliders.

    [   ]

5. Now use the same search engine to find an example of a modern aircraft (any modern airplane will do). You might search for "military aircraft" or "commercial aircraft." Draw this aircraft.

    [   ]

As you can see, in just over 100 years, aircraft design has come a long way!

Science and Technology: Student Activity Page

Inventing the Flying Machine *(cont.)*

Name: _____ Date: _____ Period: _____

Directions: Cut out the airplane pieces below to create a model airplane. Then attach a length of thread to the top of the plane and ask your teacher to hang it from the classroom ceiling.

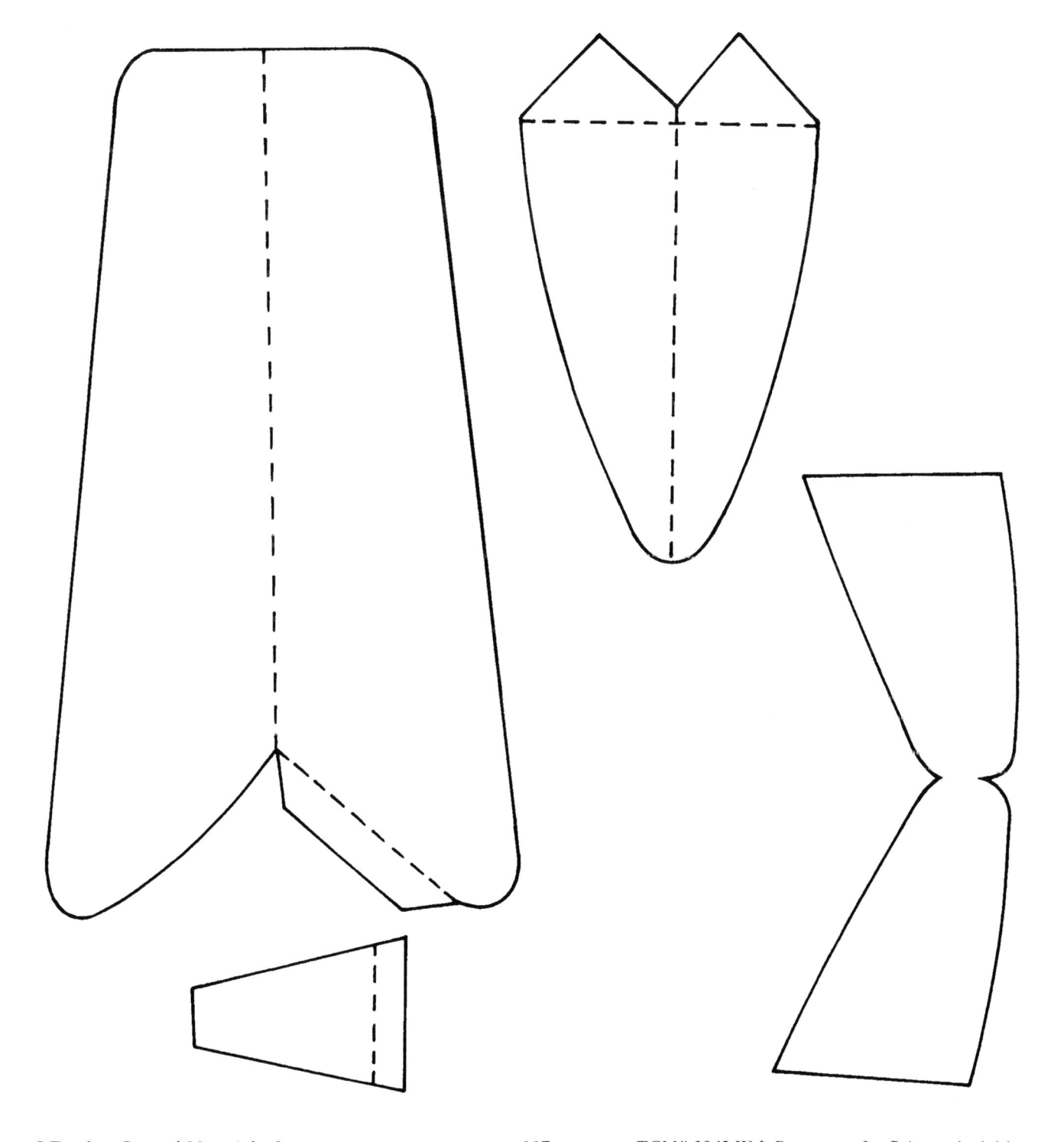

©Teacher Created Materials, Inc.   TCM# 3842 Web Resources for Science Activities

Science and Technology: Teacher Notes

Our World from Space***

NSTA Standards (5–8):

A, D, E, F

Objectives:

Students will:
- learn about man-made satellites and the uses of satellite data
- find, draw, and describe a satellite
- find and draw a satellite image of Earth
- infer what scientists can learn from the satellite image

Suggested Web Site:

http://www.teachercreated.com/books/3842

Click page 118, site 1

Alternate Web Site(s):

Satellites

http://collections.ic.gc.ca/satellites/english/

Google

http://www.google.com

Type the keywords "satellites kids" to find Web sites.

Pre-Internet Activities:

- Briefly allow students to share what they know about satellites. Prompt them with questions such as, "what are satellites?," "where are satellites?," and "what do satellites do for us?"
- Share the following with students to give them some background information:

    A satellite is any object that orbits or revolves around another object. For instance, the moon is a satellite of the earth. However, usually when we talk about satellites, we mean man-made equipment that orbits the earth. Satellites do many things for us, like relaying TV and telephone signals (cells phones are dependent upon satellites) to earth, collecting data for scientific research, tracking global weather, and serving as experiment stations.
    Satellites carry sensors, cameras, radar, and other scientific instruments to collect data about the earth. Some satellites take pictures of the earth. Because of where they are positioned out in space, satellites can take pictures of earth that can be very valuable for studying the environment and the weather.

- Provide colored pencils for your students to draw the satellite image.

Extensions:

- Discuss how gravity keeps satellites (including the moon) in orbit around the earth.
- Allow students to send a satellite image postcard to a friend or family member at
    http://gaia.hq.nasa.gov/ecard/index.cfm?kids=yes
    (Note: this Web page address was valid at the time of this book's printing.)

Science and Technology: Student Activity Page

Our World from Space (cont.)

Name: _____ Date: _____ Period: _____

Directions: Go to **http://www.teachercreated.com/books/3842**
and click page 118, site 1. Use the Web site to learn more about man-made satellites. Be sure to draw detailed pictures.

1. Find a picture of a satellite that has a description next to it. Draw the satellite here:

2. Describe this satellite. Tell what kind of satellite it is and what it does:

_____
_____
_____
_____
_____

©Teacher Created Materials, Inc.   TCM# 3842 Web Resources for Science Activities

Science and Technology: Student Activity Page

Our World from Space (cont.)

Name: _____ Date: _____ Period: _____

3. Find an image of the earth taken by a satellite and draw it here:

4. What part of earth is shown in this picture? _____
_____
_____
_____

5. What do you think scientists can learn from this picture? _____
_____
_____

Bonus question: What is "space junk" and why is it a potential problem?

Science and Technology: Teacher Notes

20th Century Inventions*

NSTA Standards (5–8):
A, E, F, G

Objectives:
Students will:
- brainstorm and list 20th century inventions
- research one 20th century invention

Suggested Web Site:
http://www.teachercreated.com/books/3842
Click page 121, site 1

Alternate Web Site(s):
Google
http://www.google.com

Pre-Internet Activities:
- Direct students to think about 20th century inventions. Most of the appliances, electronic devices, computer technology, toys, and modes of transport we use today were invented in the 20th century. Allow students to brainstorm types of 20th century inventions with a partner. Instruct students to record the inventions on the student activity sheet provided.

Extensions:
- Allow students to imagine a world without the invention they researched. How would our lives be different? Challenge students to write a creative story about how a fictional character manages to live without this invention.
- Direct students to learn about a culture or society that lives without modern inventions. Such societies might include the Amish, historical societies (before the 19th or 20th centuries), or people who live deep in the Amazon away from modern influences. How do these people accomplish everyday tasks? Who does the tasks? Ask students to imagine that they went to live with these people. What sorts of things might the students have to learn? Would the students be able to share any of their technology with these people? Would it be helpful to them?
- Challenge students to imagine future inventions. What do we need to make our lives easier? What would help people communicate better? Can students imagine inventions that might keep the earth cleaner or less polluted? Ask students to brainstorm inventions together, then assign each student the task of creating an invention on paper. Students should draw the invention, describe what it does, and explain how it would benefit people.

Science and Technology: Student Activity Page

20th Century Inventions *(cont.)*

Name: _____ Date: _____ Period: _____

Directions: Below, list as many 20th century inventions as you can think of. Remember, a 20th century invention would have been created sometime between the years 1900 and 2000.

20th Century Inventions

Science and Technology: Student Activity Page

20th Century Inventions *(cont.)*

Name: _____ Date: _____ Period: _____

Directions: Now choose one invention to research from your list. Then go to **http://www.teachercreated.com/books/3842** and click page 121, site 1. Use the search engine to find information on this invention. (Hint: type the word "invention" and then the name of the invention in the Search box. For example, if you want to research the airplane, type "invention airplane" in the search engine Search box.) First, find out when it was invented to make sure it was invented in the 20th century. Then find the rest of the information requested below.

20th Century Invention Research
Name of invention:
Year invented (it should be between 1900 and 2000):
Inventor:
What were people able to do with this invention when it was first created?
What can people do with this invention now?
Ways this invention has made people's lives easier:

Science and Technology: Teacher Notes

Science vs. Technology*

NSTA Standards (5–8):
A, E, F, G

Objectives:
Students will:
- brainstorm the concepts of science and technology
- explain how science and technology are different
- research given examples of science and technology and classify each item as an example of science or technology

Suggested Web Site:
http://www.teachercreated.com/books/3842
Click page 124, site 1

Alternate Web Site(s):
Fields of Science – MIStupid.com
http://mistupid.com/science/fields.htm

KidsClick!
http://sunsite.berkeley.edu/KidsClick!/

Pre-Internet Activities:
- Ask students to brainstorm the concepts of science and technology using the comparison chart provided. When they are finished, ask them to answer the question at the bottom of the page. You may wish to allow students to work on the question together.
- Note: you may wish to pair students for the Internet activity since students may have to infer the meanings of the words from the Web sites they view. The definitions of the words may not be explicitly stated.

Extensions:
- A Science Odyssey: You Try It: Technology at Home
  http://www.pbs.org/wgbh/aso/tryit/tech/#
  If this site is still online, be sure to have your students use it! This wonderful interactive activity shows students how technology has changed over time from 1900 through 1998.
- Think of examples of technology that your students can look up using the Internet. For each one, challenge them to figure out what problem the product solves or what need the product meets.
- Direct students to research a technology that has improved over time, such as the refrigerator, telephone, computer, calculator, etc. Ask them to find out when it was first invented, what major forms it has taken, what features have been added, and how the original form of this technology compares to today's form. Arrange for students to present their research to each other in the form of computer presentations or reports. Note that the site mentioned above, A Science Odyssey: You Try It: Technology at Home (**http://www.pbs.org/wgbh/aso/tryit/tech/#**) will help your students with this research project (if the site is still online).

Science and Technology: Student Activity Page

Science vs. Technology (cont.)

Name: _____ Date: _____ Period: _____

Directions: Brainstorm what you know about science and technology in the columns below. Write what you know about science in the Science column, and likewise, write what you know about technology in the Technology column.

Science	Technology

After you have finished brainstorming, answer the following question:

How are science and technology different?

Science and Technology: Student Activity Page

Science vs. Technology (cont.)

Name: _____ Date: _____ Period: _____

Science is the study of the natural world. Technology, on the other hand, is a solution to a problem or, a product that meets a need. The study of plants is science. The pesticides used to kills pests that eat plants are a form of technology. Science is used in the making of the pesticides, but the pesticides themselves are not science—they are a technological product.

Directions: Go to **http://www.teachercreated.com/books/3842**
and click page 124, site 1.

Use the search engine to look up the words below. To do this, type the word in the search engine Search box, view the list of results, and click the best Web site listed to view it. You may need to look at more than one Web site to determine the meaning of the word. Write what each word means in the middle column. Then, decide whether the word describes a science or a technology, and circle the appropriate choice. The first word is done for you.

Word	Definition	Is it Science or Technology? (Circle one)
Botany	The study of plants.	**Science**/Technology
Catapult		Science/Technology
Phonograph		Science/Technology
Oceanography		Science/Technology
Paleontology		Science/Technology
Laser		Science/Technology

Bonus question: Is robotics a science or a technology?

Physical Science: Teacher Notes

Electricity and Magnetism**

NSTA Standards (5–8):
A, B, F, G

Objectives:
Students will:
- identify and research a scientist involved in the study of electricity and magnetism
- learn about the principles of electricity and magnetism discovered or explored by the scientist researched

Suggested Web Site:
http://www.teachercreated.com/books/3842

Click page 127, site 1

Alternate Web Site(s):
Benjamin Franklin – Inventor

http://bensguide.gpo.gov/benfranklin/inventor.html

Science Heroes – Thomas Alva Edison

http://myhero.com/hero.asp?hero=ta_edison

Pre-Internet Activities:
- Ask students to brainstorm "What is electricity?" and "What is magnetism?" to refresh their memories of what they learned about both of these topics in K-4.
- If necessary, lead students through the process of using the Internet to identify a scientist who works or has worked in the fields of electricity and magnetism. Or you might choose to assign students to specific scientists that you have already identified (see list below).
- Discuss why it is important to record the Web site names and addresses from which students pull the information they use for their reports.

Extensions:
- Allow students to create a report or book about their scientists using *Microsoft Word, Microsoft Publisher, Microsoft PowerPoint,* or another word processing or presentation software. As an alternative, direct students to create a handmade report or book about their scientist. Components might include a hand-drawn picture of the scientist, a picture or drawing of something the scientist discovered or explored, or several pages that contain both text and illustrations.

Scientists who worked in the fields of electricity and magnetism:

Hans Christian Oersted	André-Marie Ampère	Michael Faraday	Alessandro Volta
Benjamin Franklin	George Ohm	Thomas Edison	Gaston Plante
Robert J. Van de Graaff	Nikola Tesla		

Physical Science: Student Activity Page

Electricity and Magnetism (cont.)

Name: _____ Date: _____ Period: _____

Directions: Go to **http://www.teachercreated.com/books/3842** and click page 127, site 1. Use the Web site to find a scientist who worked in the field of electricity and magnetism. To do this, try typing "scientists electricity magnetism" in the search engine Search box. Choose the best Web site from the list. Then use the information you find to answer the questions below.

Research Guide

Name of Scientist:
Birth/Death:
Discoveries, theories, and/or inventions:
Biographical information:
Other interesting notes:
Where I found my information:

Physical Science: Teacher Notes

Newton's First Law of Motion**

NSTA Standards (5–8):
A, B, G

Objectives:
Students will:
- learn about the scientist Sir Isaac Newton and some of his accomplishments
- understand Newton's First Law of Motion

Suggested Web Site:
http://www.teachercreated.com/books/3842
Click page 129, site 1

Alternate Web Site(s):
Ask Jeeves for Kids
http://www.ajkids.com

Pre-Internet Activities:
- Ask students to brainstorm the reason why one's body continues moving forward in a car that stops suddenly. Or, ask them to explain why a ball rolling along the ground will eventually come to a stop even if no one or no object stops it.

Extensions:
- Use this lesson as a launching pad into a study of motions and forces. Reinforce the ideas learned in this lesson with concrete objects such as balls. Study the concept of friction in more depth.

Explanation of Sir Isaac Newton's First Law of Motion:

Sir Isaac Newton's First Law of Motion is stated: An object at rest tends to stay at rest and *an object in motion tends to stay in motion with the same speed and in the same direction unless acted upon by an unbalanced force.*

A good way to explain the First Law of Motion is with the example of a moving car. When the driver slams the brakes in a moving car, the passengers' bodies continue to move forward until the seat belts stop them. The passengers' bodies stayed in motion with the same speed and direction as the moving car until they were acted upon by the seatbelts.

Another way to explain Newton's First Law of Motion is with the example of a ball rolling along the ground. The ball would stay in motion at the same speed and direction if it weren't for the friction the ball encounters on the ground, which slows it down until it eventually comes to a stop.

Physical Science: Student Activity Page

Newton's First Law of Motion

Name: _____ Date: _____ Period: _____

Directions: Go to **http://www.teachercreated.com/books/3842** and click page 129, site 1. Use the Web site to answer the questions below. Type "Isaac Newton" in the search engine Search box to bring up a list of sites that will help you answer the first question. For the remaining questions, type "first law of motion" in the search engine Search box. Choose the best Web site(s) from the list to help you answer the questions.

1. Who was Sir Isaac Newton and what were some of his accomplishments?

   _____
   _____
   _____
   _____
   _____

2. Write out Newton's First Law of Motion here:

   _____
   _____
   _____
   _____
   _____

3. Listen as your teacher explains the First Law of Motion. Now write an explanation of the First Law of Motion in your own words:

   _____
   _____
   _____
   _____
   _____

Physical Science: Teacher Notes

Elements**

NSTA Standards (5–8):
A, B

Objectives:
Students will:
- find and name several elements
- find the approximate number of known elements
- record other facts about elements
- complete a partially blank periodic table

Suggested Web Site:
http://www.teachercreated.com/books/3842

Click page 131, site 1

Alternate Web Site(s):
Ask Jeeves for Kids

http://www.ajkids.com

Pre-Internet Activities:
- Introduce students to elements. Elements are the building blocks of all living and non-living things. Everything is made of elements. Elements cannot be broken down into more substances by heating or chemical reactions. There are only so many elements in the entire universe, and elements found on earth are the same as those found on other planets. Examples of elements include oxygen, hydrogen, helium, iron, gold, and silver.

Extensions:
- At the time of this book's printing, Yahooligans provided a game called Element Lab on its Web site. To find it, click **Games** on the Yahooligans home page. Note: Preview the game before directing your students to play it to make sure the difficulty level is appropriate.
- Some Web sites provide interesting information on elements, such as discovery dates, discoverers, what the elements are called in different languages, and uses of the elements. Direct each student to choose a few elements to research using this site. Then have students present the information to the rest of the class.
- Move on to study compounds. Introduce the idea that compounds are made up of two or more elements bonded together. Have students investigate what elements combine to form common substances such as water (H_2O), rust (Fe_2O_3), or salt (NaCl).

Physical Science: Teacher Notes

Elements**

Name: _____ Date: _____ Period: _____

Directions: After answering the first question below, go to
http://www.teachercreated.com/books/3842
and click page 131, site 1. Use the subject directory to learn about elements. Complete the concept map shown below as you read about elements. Hint: to find Web sites about elements, it is helpfull to know that elements are a subtopic of Chemestrey, which is a subtopic of Physical Sciences. To find the elements that make up water, try using the search engine instead, and type the keywords, "elements water."

1. Your teacher described what elements are, and now it's your turn. Describe what elements are in your own words:

_____
_____
_____
_____
_____

2. Now complete the concept map shown below:

- Other facts about elements:
- Names of elements:
- **Elements**
- Approximate number of known elements: _____

©Teacher Created Materials, Inc. 132 TCM# 3842 Web Resources for Science

Physical Science: Student Activity Page

Elements *(cont.)*

Name: _____ Date: _____ Period: _____

Directions: Use the same Web site (**http://www.teachercreated.com/books/3842**, click page 131, site 1) to fill in the blank boxes in the periodic table shown below. Be sure to write each element's abbreviation and full name in the appropriate box. To find a periodic table, try typing "periodic table" in the search engine's Search box.

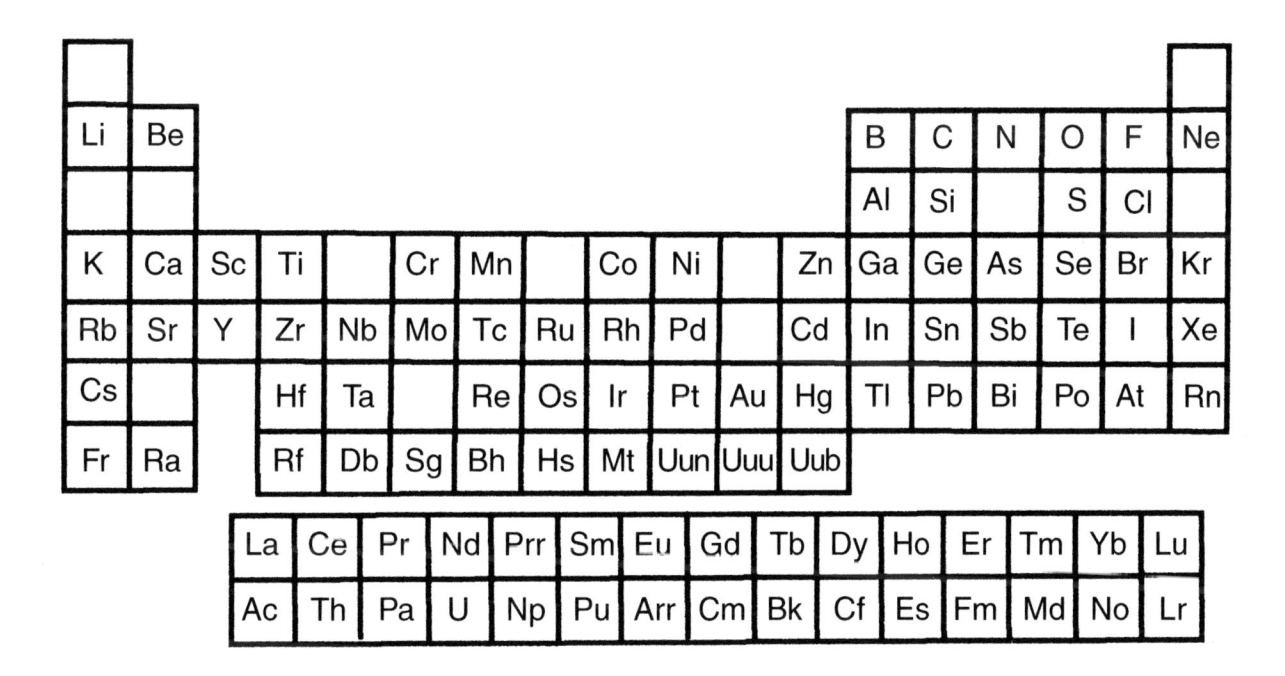

The Periodic Table

Li	Be											B	C	N	O	F	Ne
												Al	Si		S	Cl	
K	Ca	Sc	Ti		Cr	Mn		Co	Ni		Zn	Ga	Ge	As	Se	Br	Kr
Rb	Sr	Y	Zr	Nb	Mo	Tc	Ru	Rh	Pd		Cd	In	Sn	Sb	Te	I	Xe
Cs			Hf	Ta		Re	Os	Ir	Pt	Au	Hg	Tl	Pb	Bi	Po	At	Rn
Fr	Ra		Rf	Db	Sg	Bh	Hs	Mt	Uun	Uuu	Uub						

La	Ce	Pr	Nd	Prr	Sm	Eu	Gd	Tb	Dy	Ho	Er	Tm	Yb	Lu
Ac	Th	Pa	U	Np	Pu	Arr	Cm	Bk	Cf	Es	Fm	Md	No	Lr

Physical Science: Teacher Notes

Chemistry in Action**

NSTA Standards (5–8):
A, B

Objectives:
Students will:
- describe the materials, results, and chemical reaction of the bubble bomb experiment
- identify the base, acid, and gas used or created in the experiment

Suggested Web Site:
http://www.teachercreated.com/books/3842
Click page 134, site 1

Alternate Web Site(s):
Science Explorer: Bubble Bomb
http://www.exploratorium.edu/science_explorer/bubblebomb.html

CHEM4KIDS
http://www.chem4kids.com

Pre-Internet Activities:
- Study how substances can be combined and react with each other in a chemical way. Explain that chemical reactions always produce new substances. Give examples of everyday chemical reactions that students are familiar with, such as rusting, which happens when the iron in metal combines with oxygen in the air.
- Briefly explain acids and bases. Acids and bases are liquids with different amounts of certain kinds of ions. (Explain what ions are if appropriate). Scientists use a pH scale to measure how acidic or basic a liquid is. Acidic liquids have a pH between 0 and 7. Basic liquids have a pH between 7 and 14. Most everyday liquids have a neutral pH near 7. (Note: the base in this experiment is the baking soda. Students may point out that baking soda is not a liquid. Explain that baking soda becomes a base when it is combined with the water in the bag.)

Extensions:
- Allow students to create bubble bombs under your supervision in a place where clean up is easy, such as outside. Consider pairing students for easier instruction and clean up.
- Experiment with different size bags and ask students to predict what will happen with each one. If a larger bag won't explode, ask them to hypothesize why it didn't (not enough gas was produced to exceed the volume of the bag). Allow students to attempt to explode a larger bag by adding more of the ingredients.

Physical Science: Student Activity Page

Chemistry in Action *(cont.)*

Name: _____ Date: _____ Period: _____

Directions: Go to **http://www.teachercreated.com/books/3842** and click page 134, site 1. Use the search engine to find a Web site that explains the bubble bomb experiment. Be sure to use the keywords "bubble bomb experiment" for best results. Then complete the blank areas below.

The Bubble Bomb Experiment

What materials are needed?

Simplified Steps (note that if you are going to try this experiment, do it with adult supervision and follow more detailed instructions than the ones listed below).

1. Find a bag that can be sealed shut and is free of holes.
2. Fold the baking soda into a paper towel.
3. Pour vinegar and warm water into the bag.
4. Drop the baking soda square into the bag and zip it shut quickly.
5. Shake the bag and stand back!

What happened?

Physical Science: Student Activity Page

Chemistry in Action *(cont.)*

Name: _____ Date: _____ Period: _____

What was the chemical reaction?

The chemical reaction:

        was mixed with   to create the gas

◇ + ▭ → ⬯

In this experiment, an acid reacted with a base to create a gas. Which one is which?

The base was: _____

The acid was: _____

The gas was: _____

Bonus question:

Write the chemical reaction using element symbols (Hint: if the Web site does not show the experiment written this way, find the symbols by typing the name of the substance followed by "element symbols" or "chemical notation.").

_____ + _____ → _____

Physical Science: Teacher Notes

Forms of Energy***

NSTA Standards (5–8):
A, B

Objectives:
Students will:
- understand the basic definition of energy as the ability to do work
- define kinetic energy and potential energy
- give an example of kinetic energy and potential energy

Suggested Web Site:
http://www.teachercreated.com/books/3842
Click page 137, site 1

Alternate Web Site(s):
Energizing Buddies Energy Basics
http://www.energizingbuddies.cc/energy.html

Pre-Internet Activities:
- Allow students to brainstorm the concept of energy using the concept map provided.
- Discuss energy. Describe energy as the ability to do work. Energy is everywhere, not just in us. It is in the sun, in a dog, in the trees, in a car, and in water. There are two major kinds of energy: kinetic and potential energy. Tell students that they will learn about kinetic energy and potential energy today using the Internet.

Extensions:
- To reinforce what students have learned about potential and kinetic energy, give several examples of stored energy or work being done, and instruct students to label each one as an example of potential or kinetic energy. You may choose to give these examples: a rock falling off a cliff (kinetic energy), energy in food (potential energy), a cat playing with a ball of string (kinetic energy), energy in a battery (potential energy), a basketball rolling across the floor (kinetic energy), and a compressed spring (potential energy).
- Discuss energy further. Explain that energy cannot be destroyed, simply changed from one form to another. For example, a dog lying on the floor has potential energy. Once the dog gets up and moves, the potential energy has been changed to kinetic energy. Give more examples of potential energy being converted to kinetic energy.
- Many Web sites provide instructions on how to demonstrate energy concepts. Type the keywords "energy lesson plans," or "kinetic potential energy lesson" to find these instructions.

Physical Science: Student Activity Page

Forms of Energy *(cont.)*

Name: _____ Date: _____ Period: _____

Directions: Use the concept map below to brainstorm everything you know about energy.

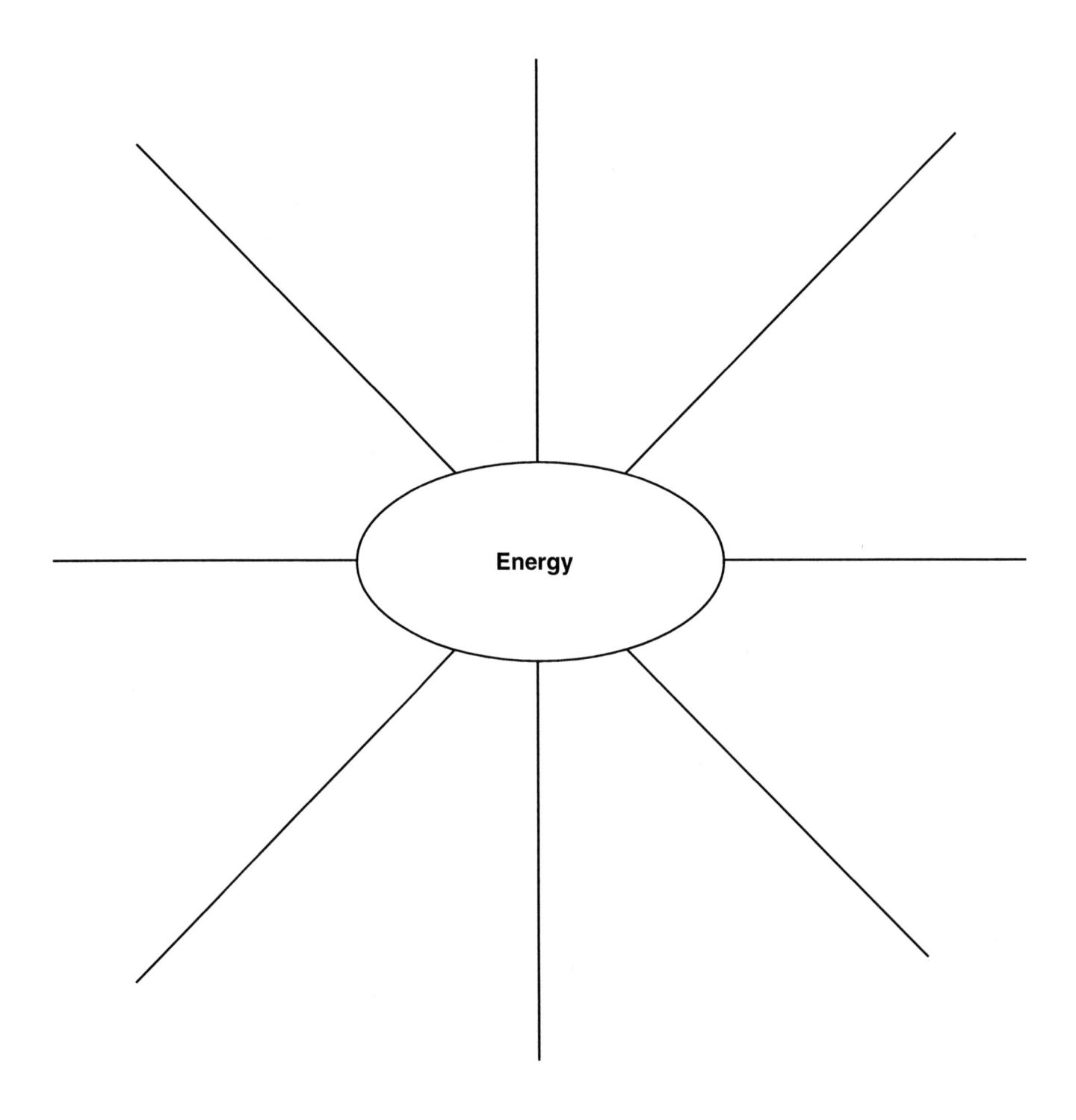

Energy

Physical Science: Student Activity Page

Forms of Energy (cont.)

Name: _____ Date: _____ Period: _____

Directions: Go to **http://www.teachercreated.com/books/3842** and click page 137, site 1. Use the search engine to find a Web site about kinetic and potential energy. Read about each form of energy, write a definition for each in your own words, and then think of an example. Hint: use the keywords "kinetic potential energy kids" to find Web sites written for students.

	Kinetic Energy	**Potential Energy**
Definition		
Example		

What are some other forms of energy? (Hint: heat is another form of energy.)

Bonus question: Label each of the forms of energy you listed above as either kinetic or potential. Be sure to check your work using the Internet.

Physical Science: Teacher Notes

A Sun's Energy*

NSTA Standards (5–8):
A, B, C, D

Objectives:
Students will:
- discover that the sun is the major source of energy for the earth
- understand that energy is transferred to earth by heat and wavelengths of light
- find and record ten additional facts about the Sun

Suggested Web Site:
http://www.teachercreated.com/books/3842

Click page 140, site 1

Alternate Web Site(s):
NASA Kids – Sun

http://kids.msfc.nasa.gov/SolarSystem/Sun/

Stanford SOLAR Center – About the Sun

http://solar-center.stanford.edu/about.html

Pre-Internet Activities:
- Discuss how the earth is dependent upon the Sun. Ask students to come up with three things the Sun makes possible on earth, then share ideas. Then point out that heat and light are forms of energy, and the Sun transfers this energy to earth. Then direct students to read the top of the A Sun's Energy student activity sheet to learn more about the Sun's energy.

Extensions:
- Challenge students to think about why there is no life (as far as we know!) on the other planets in our solar system. Why does the Sun support life on our planet but not the others? After students have thought about the question for a while, ask them to find a diagram of our solar system on the Internet. Looking at the diagram, ask them to think about how planet position relative to the Sun might affect the ability of the Sun to support life on each planet.
- Tell students that sunlight is the major source of energy for all ecosystems. Have them think alone or in groups why this is so. If they are struggling for an answer, tell them to start by thinking about plants. Plants turn sunlight into food by the process of photosynthesis. (If students are unfamiliar with this term, instruct them to look it up on the Internet.) In this way, plants are transferring energy from the sun to themselves. Then animals eat the plants and energy is transferred again. If desired, have students draw a food web illustrating this concept. Examples of food webs can be found on the Internet.

Physical Science: Student Activity Page

A Sun's Energy (cont.)

Name: _____ Date: _____ Period: _____

The Sun is the major source of energy for the earth. Without the Sun, animals, plants, and people would not survive. Green plants use the Sun's rays as energy to grow. These green plants in turn provide us with oxygen and food. The Sun also warms the oceans and evaporates the water, which in turn falls on the land in the form of rain or snow, giving us water to drink. And of course, heat from the Sun provides the earth with a climate hospitable to life.

Light and heat are the forms of energy that the Sun transfers to the earth. The Sun emits light in a range of wavelengths - visible light, infrared, and ultraviolet radiation.

Directions: Go to **http://www.teachercreated.com/books/3842** and click page 140, site 1. Use the subject directory to find a Web site about the Sun and record 10 facts on the Sun's rays below. Hint: The Sun is a subtopic of the Solar System, which is a subtopic of Astronomy and Space.

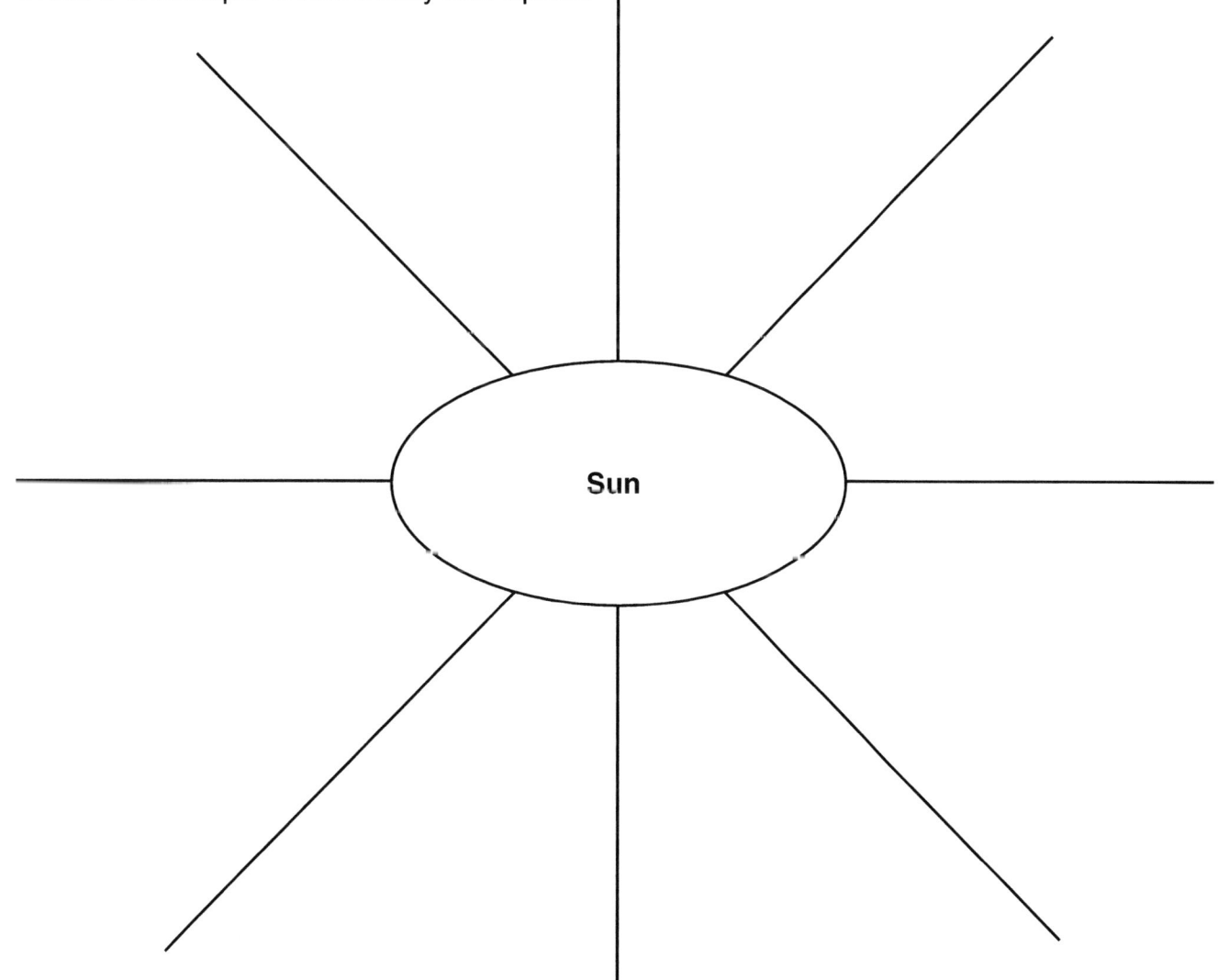

Answer Key

Pg. 11

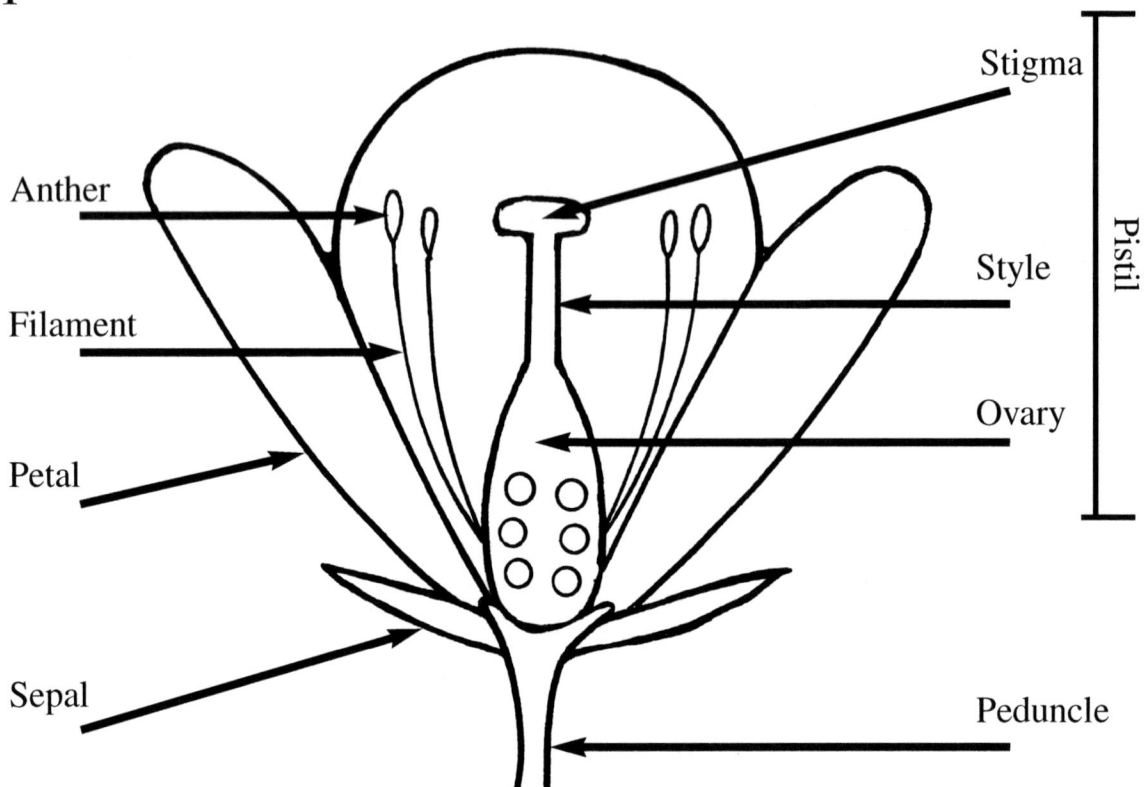

Labels: Anther, Filament, Petal, Sepal, Stigma, Style, Ovary, Pistil, Peduncle

Pg. 18

  6
  3
  5
  2
  1
  4

Pg. 32

Scientific Name	Common Name
Cereus giganteus	Saguaro cactus
Acer macrophyllum	Bigleaf Maple Tree
Citrus limon	Lemon tree
Felis cattus	Cat
Equus caballus	Horse
Rhus toxicodendron	Poison Ivy
Canis familiaris	Dog

Answer Key (cont.)

Page 50

1. An ecosystem is a group of living parts, such as plants and animals, and non-living parts, such as sunlight, air, water, and soil, that supports life.
2. A producer is an organism that can make its own food through photosynthesis. Most producers are green plants.
3. A consumer is an organism that eats other organisms.
4. A decomposer is an organism that breaks down dead organic matter into gasses like carbon and nitrogen.

Page 54

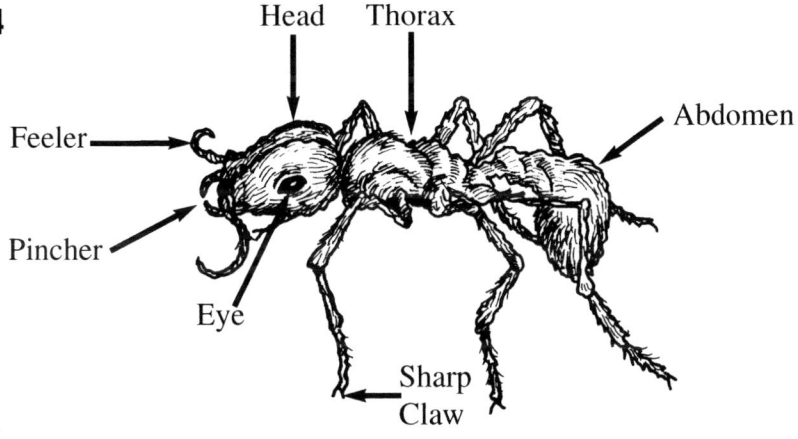

Labels on ant diagram: Head, Thorax, Abdomen, Feeler, Pincher, Eye, Sharp Claw

Page 68

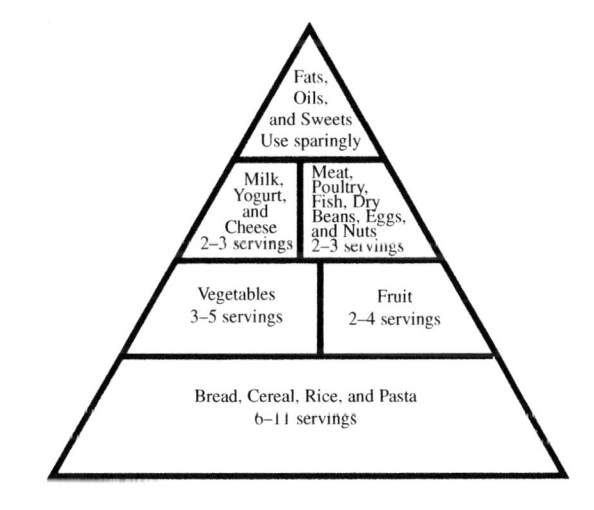

Food Pyramid:
- Fats, Oils, and Sweets — Use sparingly
- Milk, Yogurt, and Cheese — 2–3 servings
- Meat, Poultry, Fish, Dry Beans, Eggs, and Nuts — 2–3 servings
- Vegetables — 3–5 servings
- Fruit — 2–4 servings
- Bread, Cereal, Rice, and Pasta — 6–11 servings

Page. 18

3, 4, 2, 5, 1

Page 79

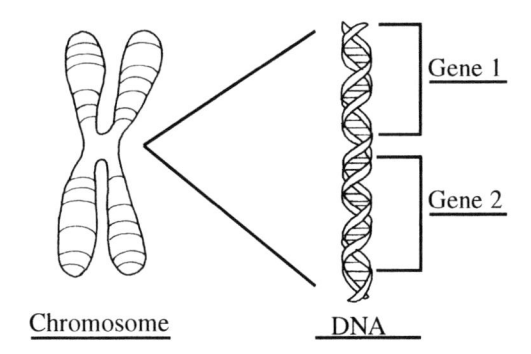

Labels: Chromosome, DNA, Gene 1, Gene 2

Answer Key (cont.)

Page 101

1, 2, 3, 4, 5, 6, 7, 8

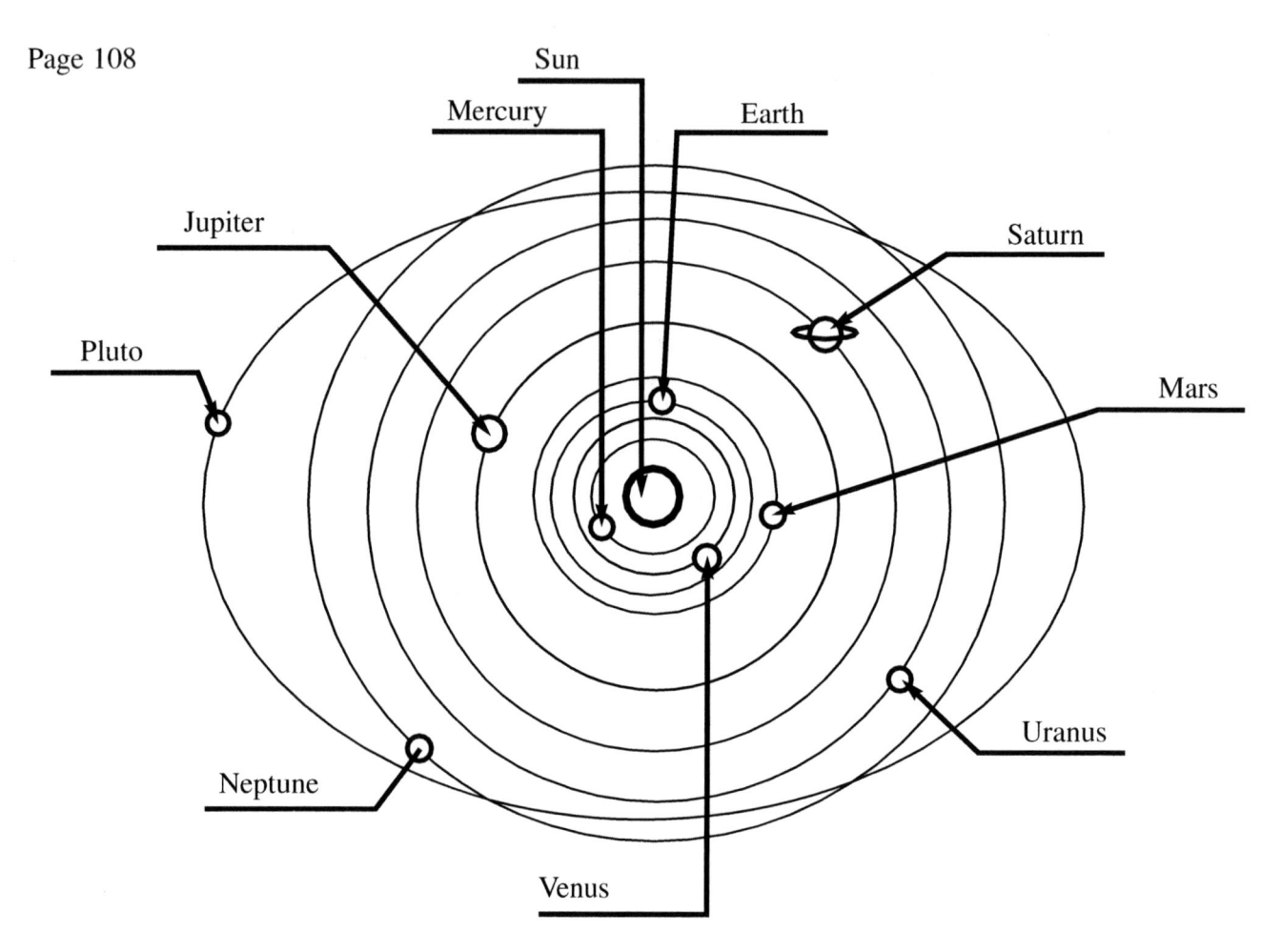

Page 108

(Diagram of solar system with labels: Sun, Mercury, Earth, Jupiter, Saturn, Pluto, Mars, Neptune, Venus, Uranus)

Page 133

(Periodic table with highlighted elements: H, Na, Mg, V, Fe, Cu, Ag, Ba, W, P, Ar, He)

H																	He
Li	Be											B	C	N	O	F	Ne
Na	Mg											Al	Si	P	S	Cl	Ar
K	Ca	Sc	Ti	V	Cr	Mn	Fe	Co	Ni	Cu	Zn	Ga	Ge	As	Se	Br	Kr
Rb	Sr	Y	Zr	Nb	Mo	Tc	Ru	Rh	Pd	Ag	Cd	In	Sn	Sb	Te	I	Xe
Cs	Ba		Hf	Ta	W	Re	Os	Ir	Pt	Au	Hg	Tl	Pb	Bi	Po	At	Rn
Fr	Ra		Rf	Db	Sg	Bh	Hs	Mt	Uun	Uuu	Uub						

La	Ce	Pr	Nd	Prr	Sm	Eu	Gd	Tb	Dy	Ho	Er	Tm	Yb	Lu
Ac	Th	Pa	U	Np	Pu	Arr	Cm	Bk	Cf	Es	Fm	Md	No	Lr